T0257152

The Ergodic Theory of Lattice Subgroups

Alexander Gorodnik
Amos Nevo

PRINCETON UNIVERSITY PRESS

PRINCETON AND OXFORD

2010

Published by Princeton University Press, 41 William Street, Princeton, New Jersey 08540

In the United Kingdom: Princeton University Press, 6 Oxford Street, Woodstock, Oxfordshire OX20 1TW

Library of Congress Cataloging-in-Publication Data

Gorodnik, Alexander, 1975–

  The ergodic theory of lattice subgroups / Alexander Gorodnik and Amos Nevo.

    p. cm. — (Annals of mathematics studies ; no. 172)

  Includes bibliographical references and index.

  ISBN 978-0-691-14184-8 (hardcover) — ISBN 978-0-691-14185-5 (pbk)

1. Ergodic theory. 2. Lie groups. 3. Lattice theory. 4. Harmonic analysis. 5. Dynamics. I. Nevo, Amos, 1960– II. Title.

  QA313.G67 2010

  515′.48—dc22

                                                                    2009003729

British Library Cataloging-in-Publication Data is available

Printed on acid-free paper. ⊗

press.princeton.edu

Printed in the United States of America

10 9 8 7 6 5 4 3 2 1

Annals of Mathematics Studies
Number 172

Contents

Chapter 5. Proof of ergodic theorems for S-algebraic groups 47
5.1 Iwasawa groups and spectral estimates 47
5.2 Ergodic theorems in the presence of a spectral gap 50
5.3 Ergodic theorems in the absence of a spectral gap, I 56
5.4 Ergodic theorems in the absence of a spectral gap, II 57
5.5 Ergodic theorems in the absence of a spectral gap, III 60
5.6 The invariance principle and stability of admissible averages 67

Chapter 6. Proof of ergodic theorems for lattice subgroups 71
6.1 Induced action 71
6.2 Reduction theorems 74
6.3 Strong maximal inequality 75
6.4 Mean ergodic theorem 78
6.5 Pointwise ergodic theorem 83
6.6 Exponential mean ergodic theorem 84
6.7 Exponential strong maximal inequality 87
6.8 Completion of the proofs 90
6.9 Equidistribution in isometric actions 91

Chapter 7. Volume estimates and volume regularity 93
7.1 Admissibility of standard averages 93
7.2 Convolution arguments 98
7.3 Admissible, well-balanced, and boundary-regular families 101
7.4 Admissible sets on principal homogeneous spaces 105
7.5 Tauberian arguments and Hölder continuity 107

Chapter 8. Comments and complements 113
8.1 Lattice point–counting with explicit error term 113
8.2 Exponentially fast convergence versus equidistribution 115
8.3 Remark about balanced sets 116

Bibliography 117

Index 121

Preface

0.1 MAIN OBJECTIVES

Let G be a locally compact second countable (lcsc) group and let $\Gamma \subset G$ be a discrete lattice subgroup. The present volume is devoted to the study of the following four problems in ergodic theory and analysis that present themselves in this context, namely:

  I. Prove ergodic theorems for general families of averages on G,

 II. Solve the lattice point–counting problem (with explicit error term) for general domains in G,

III. Prove ergodic theorems for arbitrary measure-preserving actions of the lattice subgroup Γ,

 IV. Establish equidistribution results for isometric actions of the lattice subgroup Γ.

We will give a complete solution to these problems for fairly general averages and domains in all noncompact semisimple algebraic groups over arbitrary local fields and any of their discrete lattice subgroups. Our results also apply to lattices in products of such groups and thus to all semisimple S-algebraic groups and their lattices. In fact, many of our arguments apply in greater generality still and serve as a general template for proving ergodic theorems for actions of a general lcsc group G and of a lattice subgroup Γ. We will elaborate on that further in our discussion below.

Let us proceed to give one concrete example of our results. Consider the lattices $\Gamma = \mathrm{SL}_n(\mathbb{Z}) \subset \mathrm{SL}_n(\mathbb{R}) = G$ and let $\sigma : \mathrm{SL}_n(\mathbb{R}) \to \mathrm{SL}_N(\mathbb{R})$ be any faithful linear representation. Let m_G be a Haar measure on G. Given any norm on $\mathrm{M}_N(\mathbb{R})$, let $G_t = \{g \in G \, ; \, \log \|\sigma(g)\| < t\}$. We will establish, in particular, the following ergodic theorems for actions of the group G and the lattice Γ (which we state for simplicity in less than their full generality).

Theorem A.

  1. In any ergodic measure-preserving action of G on a probability space (Y, ν), for $f \in L^2(Y)$,

$$\frac{1}{m_G(G_t)} \int_{g \in G_t} f(g^{-1}y)dm_G(g) \longrightarrow \int_Y f d\nu \quad as \; t \to \infty$$

  pointwise almost everywhere and in the L^2-norm.

2. *Furthermore, if $n \geq 3$, or $n = 2$ and the action has a spectral gap, then for almost every point the convergence to the limit is exponentially fast:*

$$\left| \frac{1}{m_G(G_t)} \int_{g \in G_t} f(g^{-1}y) - \int_Y f \, dv \right| \leq C(y, f) e^{-\theta t},$$

with $\theta > 0$ an explicit function of the spectral gap and the rate of volume growth of G_t.

3. *The number of lattice points in G_t satisfies*

$$\frac{|\Gamma \cap G_t|}{\text{vol } G_t} = (\text{vol } G/\Gamma)^{-1} + O\left(e^{-\theta' t}\right),$$

with $\theta' > 0$ an explicit function of θ.

4. *In any ergodic measure-preserving action of Γ on a probability space (X, μ), for $f \in L^2(X)$, setting $\Gamma_t = G_t \cap \Gamma$:*

$$\frac{1}{|\Gamma_t|} \sum_{\gamma \in \Gamma_t} f(\gamma^{-1}x) \longrightarrow \int_X f \, d\mu \quad \text{as } t \to \infty$$

pointwise almost everywhere and in the L^2-norm.

5. *Furthermore, if $n \geq 3$, or $n = 2$ and the action has a spectral gap, then for almost every point the convergence to the limit is exponentially fast:*

$$\left| \frac{1}{|\Gamma_t|} \sum_{\gamma \in \Gamma_t} f(\gamma^{-1}x) - \int_X f \, d\mu \right| \leq C(x, f) e^{-\theta'' t},$$

with $\theta'' > 0$ an explicit function of the spectral gap and the rate of volume growth of G_t.

6. *If the action of Γ is an isometric action on a compact metric space, with invariant ergodic probability measure of full support (e.g., a profinite completion of Γ), then the convergence holds for all points and is uniform, provided that f is continuous.*

In the following two sections, we will make some comments aimed at putting the foregoing results in perspective and give a brief outline of our approach to their proof.

0.2 ERGODIC THEORY AND AMENABLE GROUPS

Classical ergodic theory, developed by Poincare, von Neumann, and Birkhoff, studies a measurable space $(X, \mathcal{B})$ equipped with a probability measure μ and an action T_t of a one-parameter group that preserves the measure. A basic problem is to understand the statistical distribution of the orbits $\{T_t x\}_{t \in \mathbb{R}}$ for μ-generic points

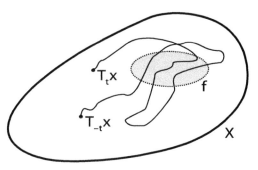

Figure 1 Distribution of orbits

$x \in X$. Given a measurable function $f : X \to \mathbb{R}$ and an initial point $x \in X$, one considers the averaging operator

$$(A_t f)(x) = \frac{1}{2t} \int_{-t}^{t} f(T_s x) \, ds,$$

which samples the values of f on part of the trajectory $\{T_t x\}_{t \in \mathbb{R}}$ and averages them with uniform distribution on the interval.

As $t \to \infty$, the quantity $(A_t f)(x)$ gives an increasingly accurate measurement of the distribution of the trajectory starting at x, so that computing $\lim_{t \to \infty}(A_t f)(x)$ is of fundamental importance. This problem was solved by von Neumann and Birkhoff who showed that under a suitable irreducibility assumption (namely, er-godicity),

$$\lim_{t \to \infty} (A_t f)(x) = \int_X f \, d\mu. \tag{0.1}$$

In von Neumann's formulation, the convergence holds in L^2-norm, and in Birk-hoff's, μ–almost everywhere. When the action is by isometries of a compact metric space, an extension of Weyl's equidistribution theorem asserts that the convergence is uniform and holds for every initial point x, provided f is continuous.

Classical examples of dynamical systems describe the evolution of physical sys-tems in time and are naturally described by actions of the one-parameter group $\mathbb{R}$. In a more general formulation, the basic goal of ergodic theory is to analyze the sta-tistical distribution of group orbits in dynamical systems consisting of a probability space $(X, \mathcal{B}, \mu)$ and a locally compact group G acting on X by measure-preserving transformations. Fixing a (right-invariant) Haar measure m_G on G and an increas-ing family of compact subsets G_t of G with $m_G(G_t) \to \infty$, one considers the averaging operators

$$(A_t f)(x) = \frac{1}{m_G(G_t)} \int_{G_t} f(g^{-1} x) dm_G(g),$$

which constitute a sampling of the values of f along the orbit $G \cdot x$ with respect to (w.r.t.) the Haar-uniform probability measure on G_t. Under the assumption of ergodicity of the G-action (i.e., that any measurable set invariant under G is either null or conull), one seeks to show that the samples $A_t f(x)$ converge to the average of f on the space X (w.r.t. the measure μ) as $t \to \infty$, namely, to $\int_X f d\mu$.

Taking the classical ergodic theorems of von Neumann, Birkhoff, and Weyl as a guide, one should aim to establish, for families of sets G_t which are as general as possible, the following:

1. *Mean ergodic theorem:* $\|A_t f - \int_X f d\mu\|_{L^2} \to 0$ as $t \to \infty$,

2. *Pointwise ergodic theorem:* $A_t f(x) \to \int_X f d\mu$ as $t \to \infty$ for almost every initial point $x \in X$,

3. *Equidistribution:* $\max_{x \in X} |A_t f(x) - \int_X f d\mu| \to 0$ as $t \to \infty$, when the action is by isometries of a compact metric space, μ has full support, and the function f is continuous.

Naturally, one expects that some conditions must be imposed on the sets in the family G_t beyond $m_G(G_t) \to \infty$. Traditionally, most of the effort has concentrated on the assumption that the sets G_t are asymptotically (right-) invariant under group translation, namely,

$$m_G(G_t \triangle G_t g) = o(m_G(G_t)) \quad \text{as } t \to \infty,$$

uniformly for g in a compact set. Such an asymptotically invariant family is said to satisfy the Følner property, which thus generalizes the fundamental almost invariance property of the intervals $[-t, t]$ on $\mathbb{R}$ or $\mathbb{Z}$. We note that the existence of such a family is equivalent to amenability of the underlying group and that furthermore even in the setting of amenable groups, natural averaging sets such as balls with respect to an invariant metric need not be asymptotically invariant.

Assuming the Følner property, a variation on a classical argument of Riesz easily implies that the averages converge in norm. The problem of pointwise convergence is much more challenging. Following Birkhoff, it was considered for $\mathbb{R}^n$-actions by Wiener, and for groups of polynomial volume growth by Calderon. Their analysis introduced several fundamental techniques, including the covering argument on the acting group, the volume-doubling condition, and the transfer principle using asymptotically invariant sets. These ideas were developed further by Herz and Coifman-Weiss, Emerson-Greenleaf, Templeman, Shulman, and E. Lindenstrauss and have culminated in the proof of the pointwise ergodic theorem for ball averages on general groups of polynomial volume growth and the proof of the pointwise ergodic theorem for tempered Følner sets on general amenable groups. We refer the reader to [N6] for an up-to-date survey and the relevant references.

0.3 ERGODIC THEORY AND NONAMENABLE GROUPS

One of the aims of the present monograph is to expand the scope of ergodic theory and develop a systematic general approach that can be applied to prove ergodic theorems for nonamenable groups. To elucidate our approach, let us comment briefly on the results stated in Theorem A and the techniques involved in their proofs.

Theorem A(1) for the Haar-uniform averages β_t on the sets $G_t \subset G$ has the form of a traditional ergodic theorem stating the pointwise convergence of the time averages $A_t f(x)$ to the space averages $\int_X f \, d\mu$. However, while the results we state are analogous to Wiener's ergodic theorem for ball averages in actions of $\mathbb{R}^n$ or $\mathbb{Z}^n$ (and more generally to the ergodic theorem for groups of polynomial growth; see [N6]), the methods we employ are necesssarily completely different. In the absence of a Følner sequence, covering arguments, and the transfer principle for amenable groups, we will start with the methods developed for radial averages on semisimple Lie groups in [N2], [N3], [NS2], and [N5] and develop them further in several directions. Spectral methods associated with square functions and Sobolev arguments play a crucial role, as do geometric comparison arguments on the associated symmetric space.

Theorem A(2) for the averages β_t has a decidedly nontraditional form and displays a fundamental feature of nonamenable ergodic theory which does not arise in the ergodic theory of amenable groups. Indeed, the result states that there is a uniform rate of convergence for the averages β_t for all ergodic actions of the group $G = \mathrm{SL}_n(\mathbb{R})$ when $n \geq 3$. This phenomenon has its roots in the spectral theory of the groups involved and was first developed for radial averages on semisimple groups in [MNS]. To handle the general case we will expand and further develop the methods in [MNS] and [N4], which rely on spectral considerations and spectral estimates that will play a crucial role in our analysis.

Parts 4 and 5 of Theorem A state ergodic theorems for the uniform averages λ_t on the lattice points in $\Gamma_t = \Gamma \cap G_t$ analogous to those satisfied by β_t. In order to prove these results we will develop a completely general method to prove ergodic theorems for λ_t in actions of a lattice subgroup, provided that the corresponding ergodic theorems are satisfied for β_t in actions of G. Note that direct spectral arguments in the case of lattice subgroups cannot possibly be used since the unitary representations of a lattice are unclassifiable. Instead, we must rely on precise geometric comparison arguments to reduce the analysis of averages on Γ to that of averages on G. The regularity properties of the sets G_t therefore assume an indispensable role.

An important aspect of our results is that they hold for the uniform averages supported on general families of sets G_t satisfying only a mild regularity assumption we call *admissibility*. Roughly speaking, admissibility amounts to the Lipschitz continuity of the sets G_t under small perturbations:

$$m_G(\mathcal{O}_\varepsilon \cdot G_t \cdot \mathcal{O}_\varepsilon - G_t) = O(\varepsilon \, m_G(G_t)) \quad \text{as } t \to \infty,$$

where $\{\mathcal{O}_\varepsilon\}$ is a basis of neighborhoods of identity (see Definitions 1.1 and 3.10 below for the precise formulation).

We emphasize that families G_t defined by natural geometric structures on the group G (e.g., balls with respect to an invariant Riemannian metric or a norm in a linear representation) are indeed admissible. The admissibility condition can be viewed as a natural substitute for the Følner property of amenable groups. For instance, we establish that admissibility implies an invariance principle, namely: pointwise convergence of $A_t f(x)$ holds on a G-invariant set of initial points $x \in X$. The invariance principle will play a crucial role in the proof of ergodic theorems for lattice subgroups, and in the amenable case follows easily from the Følner property.

We remark that since spectral methods play a very prominent role in our discussion, the stability and regularity properties of β_t become one of the main technical tools. They allow an effective reduction of the analysis of the discrete averages λ_t on Γ to that of the absolutely continuous averages β_t on G. But apart from that, they are also crucial in developing the square function and Sobolev arguments we will use in our analysis. In effect, we can say that the regularity of β_t combined with spectral theory compensates for the failure of the classical large-scale methods of amenable ergodic theory, associated with polynomial volume growth, covering argument and the transfer principle. We have thus been motivated to establish such regularity results in considerable generality, allowing for a diverse array of families G_t, arising from a variety of distance functions. This problem is quite demanding technically and requires the use of a broad range of techniques. The verification of admissibility and the discussion of volume regularity issues will be addressed in Chapter 7. The utility of establishing the results for a diverse set of distance functions is demonstrated by their appearance in the natural counting problems discussed in Chapter 2.

Theorem A(3) gives a new quantitative lattice point–counting result for the domains G_t. In fact, we will give a complete solution to the lattice point–counting problem for every semisimple S-algebraic group G and a discrete subgroup Γ with finite covolume, in an admissible family of domains, in the form

$$|\Gamma \cap G_t| = \frac{m_G(G_t)}{m_G(G/\Gamma)} + O\left(m_G(G_t)^{\theta'}\right),$$

with the parameter $\theta' \in (\frac{1}{2}, 1)$ given explicitly in terms of the spectral gap.

Our analysis will demonstrate the general principle that if the Haar-uniform averages β_t supported on the domains G_t satisfy a quantitative mean ergodic theorem in their action on the homogeneous space G/Γ, then mild regularity conditions on the domains suffice to give a quantitative solution to the lattice point–counting problem in the domains. We remark that this principle holds in great generality and yields a diverse array of quantitative counting results, and we refer the reader to [GN] where they are developed systematically. The error term produced by our method matches or exceeds the best results currently available in the literature in the case of simple higher real rank groups and holds in much greater generality. Another significant aspect of this principle is that the ergodic theoretic approach, being valid for all probability measure–preserving actions simultaneously, implies that the error terms in the counting problem hold uniformly for all lattices satisfying the same spectral gap condition. In particular, such a uniform error estimate holds for all congruence subgroups of an arithmetic lattice, a result which is established and utilized in [NeSa] for the purpose of producing an affine linear sieve for counting matrices with (almost) prime entries.

Finally, Theorem A(6) states an equidistribution result for actions of the lattice. Note that convergence in the ergodic theorem holds only for almost every initial condition, and this result cannot be improved in general. However, it is crucial for many applications in diophantine approximation and number theory to establish convergence for a given specific initial condition. This is indeed the case for isometric actions, and we show that in this case the averages λ_t converge for every initial condition and convergence is uniform. As a by-product, we can conclude that the action of the lattice is uniquely ergodic.

Organization of the book

Since the statement of our results in full generality requires a significant amount of preparation and notation, we start in Chapter 1 by considering the essential case of semisimple Lie groups, and in Chapter 2 we give a number of examples of ergodic theorems and applications to the lattice point–counting problem. In Chapter 3 we set up notation and basic tools which will be used in the rest of our discussion, and in Chapter 4 we state our results in full generality and give an extended overview of the proofs. As already noted, the proofs of ergodic theorems for S-algebraic groups are quite different depending on whether the action has a spectral gap or not, but both depend on spectral theory. In Chapter 5 we begin with a discussion of spectral results and then give a proof of the ergodic theorem for G-actions in the presence of a spectral gap, followed by a proof in the absence of a spectral gap. In Chapter 6 we develop general reduction arguments to reduce the ergodic theorems for the averages on the lattice to the ergodic theorems for the averages on the ambient group. These arguments prove the ergodic theorems for lattice actions and also establish estimates of the number of lattice points in admissible sets. Chapter 7 is devoted to the study of the volume regularity properties for naturally defined families of domains. In particular, we prove that the sets defined by invariant Riemannian metrics or linear norms are admissible. Finally, in Chapter 8 we include several comments, including an explicit formula for the error term in the lattice point–counting problem and a remark on the connection between quantitative ergodic theorems and equidistribution results.

Acknowledgments

A. Gorodnik was supported in part by NSF Grants DMS-0400631 and DMS-0654413 and a RCUK fellowship, and A. Nevo was supported in part by the Institute for Advanced Study at Princeton University and ISF grant 975/05.

Chapter One

Main results: Semisimple Lie groups case

The present chapter is devoted to describing the main results in the case of connected semisimple Lie groups, which is fundamental in what follows.

1.1 ADMISSIBLE SETS

We start by introducing the notion of admissibility, which describes the families of averaging sets G_t that will be the subject of our analysis.

Let G be a connected semisimple Lie group with finite center and no nontrivial compact factors. Fix a left-invariant Riemannian metric on G and denote the associated invariant distance by d and the Haar invariant measure by m_G. Let

$$\mathcal{O}_\varepsilon = \{g \in G : d(g, e) < \varepsilon\}.$$

Definition 1.1. An increasing family of bounded Borel subsets G_t, $t > 0$, of G is called *admissible* if there exist $c > 0$ and $t_0 > 0$, $\varepsilon_0 > 0$, such that for all $t \geq t_0$ and $0 < \varepsilon \leq \varepsilon_0$,

$$\mathcal{O}_\varepsilon \cdot G_t \cdot \mathcal{O}_\varepsilon \subset G_{t+c\varepsilon}, \tag{1.1}$$

$$m_G(G_{t+\varepsilon}) \leq (1 + c\varepsilon) \cdot m_G(G_t). \tag{1.2}$$

Let us briefly note the following facts (see Proposition 3.14, Chapter 7, and Proposition 5.24 for the respective proofs).

1. Admissibility is independent of the Riemannian metric chosen to define it.

2. Many of the natural families of sets in G are admissible. In particular, the radial sets G_t projecting to the Cartan-Killing Riemannian balls on the symmetric space are admissible. Furthermore, the sets $\{g \; ; \; \log \|\tau(g)\| < t\}$, where τ is a faithful linear representation, are also admissible, for any choice of linear norm $\|\cdot\|$.

3. Admissibility is invariant under translations; namely, if G_t is admissible, so is $gG_t h$, for any fixed $g, h \in G$.

Later on we will consider the corresponding *Hölder admissibility* condition, which will also play an important role.

1.2 ERGODIC THEOREMS ON SEMISIMPLE LIE GROUPS

We define β_t to be the probability measures on G obtained as the restriction of the Haar measure to G_t, normalized by $m_G(G_t)$.

The averaging operators associated to the measures β_t when G acts by measure-preserving transformations of a standard Borel probability space (X, μ) are given by

$$\pi(\beta_t)f(x) = \frac{1}{m_G(G_t)} \int_{G_t} f(g^{-1}x)dm_G(g).$$

To state our main results, we introduce the following notation (see Chapter 3 for a detailed discussion).

1. The family β_t (and G_t) will be called *(left-) radial* if it is invariant under (left-) multiplication by some fixed maximal compact subgroup K, for all sufficiently large t. *Standard averages* are those defined in Definition 3.19.

2. The action is called *irreducible* if every noncompact simple factor acts ergodically.

3. The action is said to have a strong spectral gap if each simple factor has a spectral gap, namely, admits no asymptotically invariant sequence of unit vectors (see §3.6 for a full discussion).

4. The sets G_t (and the averages β_t) will be called *balanced* if for every nontrivial direct product decomposition $G = G(1)G(2)$ and every compact subset $Q \subset G(1)$, $\beta_t(QG(2)) \to 0$. G_t will be called *well balanced* if the convergence is at a specific rate (see §3.5 for a full discussion).

Our first main result is the following pointwise ergodic theorem for admissible averages on semisimple Lie groups.

Theorem 1.2. Pointwise ergodic theorems for admissible averages. *Let G be a connected semisimple Lie group with finite center and no nontrivial compact factors. Let (X, μ) be a standard probability Borel space with a measure-preserving ergodic action of G. Assume that G_t is an admissible family.*

1. *Assume that β_t is left-radial. If the action is irreducible, then β_t satisfies the pointwise ergodic theorem in $L^p(X)$, $1 < p < \infty$; namely, for every $f \in L^p(X)$ and for almost every $x \in X$,*

$$\lim_{t \to \infty} \pi(\beta_t)f(x) = \int_X f d\mu.$$

The conclusion also holds in reducible actions of G, provided the averages are standard, well balanced, and boundary-regular (see §3.4 and §3.5 for definitions).

2. *If the action has a strong spectral gap, then β_t converges to the ergodic mean almost surely exponentially fast; namely, for every $f \in L^p(X)$, $1 < p < \infty$, and almost all $x \in X$,*

$$\left| \pi(\beta_t) f(x) - \int_X f d\mu \right| \leq C_p(f, x) e^{-\theta_p t},$$

where $\theta_p > 0$ depends explicitly on the spectral gap (and the family G_t).

The conclusion also holds in any action of G with a spectral gap, provided the averages satisfy the additional necessary condition of being well balanced (see §3.5 and §3.7 for definitions).

We note that there are many natural examples of averages for which the conclusions of Theorem 1.2 hold. Previously, it has been established for the Haar-uniform averages on Riemannian balls in [N3], [N4], [NS2], and [MNS]. The fact that the conditions required in Theorem 1.2 are satisfied by much more general families is demonstrated in Theorems 3.15 and 3.18 below.

Regarding Theorem 1.2(1), we remark that the proof of pointwise convergence in the case of reducible actions without a spectral gap is quite involved, and we have thus assumed in that case that the averages are standard, well balanced, and boundary-regular to make the analysis tractable. However, the reducible case will be absolutely indispensable for us below since we will induce actions of a lattice subgroup to actions of G, and these may be reducible.

Regarding Theorem 1.2(2), we note that θ_p depends explicitly on the spectral gap of the action and on natural geometric parameters of G_t, and we refer the reader to §5.2 for a full discussion including a formula for a lower bound. Furthermore, Hölder admissibility is sufficient for this part, as we shall see below.

Let us now formulate the following invariance principle for ergodic actions of G, which will play an important role below in the derivation of pointwise ergodic theorems for lattices.

Theorem 1.3. Invariance principle. *Let G, (X, μ) be as in Theorem 1.2 and let G_t be an admissible family. Then for any given measurable function f on X with $f \in L^p(X)$, the set where pointwise convergence to the ergodic mean holds; namely,*

$$\left\{ x \in X \, ; \, \lim_{t \to \infty} \frac{1}{m_G(G_t)} \int_{G_t} f(g^{-1}x) dm_G(g) = \int_X f d\mu \right\}$$

contains a G-invariant set of full measure.

We note that G is a nonamenable group and that the sets G_t are not asymptotically invariant under translations (namely, do not have the Følner property). Thus the conclusion of Theorem 1.3 is not obvious, even in the case where X is a homogeneous G-action. The special case where $G = \mathrm{SO}^0(n, 1)$ and β_t are the bi-K-invariant averages lifted from ball averages on hyperbolic space $\mathbb{H}^n$ was considered earlier in [BR].

One of our applications of ergodic theorems on G is an equidistribution theorem for isometric actions of a lattice subgroup. The proof of the latter result actually depends only on the *mean* ergodic theorem for β_t, which requires less stringent conditions than the pointwise theorem. Because of its significance later on, we formulate separately the following.

Theorem 1.4. Mean ergodic theorems for admissible averages. *Let G and (X, μ) be as in Theorem 1.2 and let G_t be an admissible family.*

1. *If the action is irreducible, or the family G_t is balanced, then*

$$\lim_{t \to \infty} \left\| \pi(\beta_t)f - \int_X f d\mu \right\|_{L^p(X)} = 0, \quad f \in L^p, \ 1 \leq p < \infty.$$

2. *If the action has a strong spectral gap, or a spectral gap and the averages are well balanced, then*

$$\left\| \pi(\beta_t)f - \int_X f d\mu \right\|_{L^p(X)} \leq B_p e^{-\theta_p t}, \quad f \in L^p, \ 1 < p < \infty,$$

for the same $\theta_p > 0$ as in Theorem 1.2(2).

We remark that the mean ergodic theorem actually holds under much more general conditions, and we refer the reader to [GN] for further discussion and applications of this fact.

1.3 THE LATTICE POINT–COUNTING PROBLEM
IN ADMISSIBLE DOMAINS

Now let $\Gamma \subset G$ be any lattice subgroup; the lattice point–counting problem is to determine the number of lattice points in the domains G_t. Its ideal solution calls for evaluating the main term in the asymptotic expansion, establishing the existence of the limit, and estimating explicitly the error term. Our second main result gives a complete solution to this problem for all lattices and all families of admissible domains in connected semisimple Lie groups. The proof we give below will establish the general principle that a mean ergodic theorem in $L^2(G/\Gamma)$ for the averages β_t (with an explicit rate of convergence) implies a solution to the Γ-lattice point–counting problem in the admissible domains G_t (with an explicit estimate of the error term). In fact, this principle applies to lattices in general lcsc groups, and we will discuss this further below.

Assuming that G is connected semisimple with finite center and no nontrivial compact factors, we note that in this case the main term in the lattice count (namely, part 1 of the following theorem) was established in [Ba] (for uniform lattices), in [DRS] (for balls w.r.t. a norm), and in [EM] (in general). Error terms were considered for rotation-invariant norms in [DRS] and for more general norms recently in [Ma]. For a comparison of part 2 of the following theorem with these results, see Chapter 2.

Theorem 1.5. Counting lattice points in admissible domains. *Let G be a connected semisimple Lie group with finite center and no nontrivial compact factors. Let G_t be an admissible family of sets and let Γ be any lattice subgroup. Normalize the Haar measure m_G to assign measure 1 to a fundamental domain of Γ in G.*

1. *If Γ is an irreducible lattice, or the sets G_t are balanced, then*

$$\lim_{t\to\infty} \frac{|\Gamma \cap G_t|}{m_G(G_t)} = 1.$$

2. *If $(G/\Gamma, m_{G/\Gamma})$ has a strong spectral gap, or the sets G_t are well balanced, then, for all $\varepsilon > 0$,*

$$\frac{|\Gamma \cap G_t|}{m_G(G_t)} = 1 + O_\varepsilon \left(\exp \left(\frac{-t(\theta - \varepsilon)}{\dim G + 1} \right) \right),$$

where $\theta > 0$ depends on G_t and the spectral gap in G/Γ via

$$\theta = \liminf_{t\to\infty} -\frac{1}{t} \log \left\| \pi_{G/\Gamma}(\beta_t) \right\|_{L_0^2(G/\Gamma)}.$$

Remark 1.6.

1. Recall that the G-action on $(G/\Gamma, m_{G/\Gamma})$ is irreducible if and only if Γ is an irreducible lattice in G, namely, the projection of Γ to every nontrivial factor of G is a dense subgroup. If the lattice is reducible, a strong spectral gap will certainly not hold.

2. The G-action on G/Γ always has a spectral gap. If G has no nontrivial compact factors and the lattice Γ is irreducible, then it has a strong spectral gap (see §3.7 for more details). Whether this is true when G has compact factors seems to be an open problem.

3. When the action has a strong spectral gap, the parameter θ can be given explicitly in terms of the rate of volume growth of the sets G_t and the size of the gap—see §5.2.2.2 and Corollary 8.1.

4. Note that under the normalization of m_G given in Theorem 1.5, if $\Delta \subset \Gamma$ is a subgroup of finite index, then

$$\lim_{t\to\infty} \frac{|\Delta \cap G_t|}{m_G(G_t)} = \frac{1}{[\Gamma : \Delta]}.$$

Finally, we remark that the condition of admissibility is absolutely crucial in obtaining pointwise ergodic theorems for G, and thus also for Γ, when the action does not have a spectral gap. When the action has a spectral gap, Hölder admissibility is sufficient. However, lattice point–counting results, quantitative or not, hold in much greater generality. Namely, they hold for families that satisfy the weaker condition $m_G(\mathcal{O}_\varepsilon G_t \mathcal{O}_\varepsilon) \leq (1 + c\varepsilon) m_G(G_t)$, which amounts to a quantitative version of the well-roundedness condition in [DRS] and [EM]. This generalization is discussed systematically in [GN], where several applications, including those to quantitative counting of lattice points in sectors, on symmetric varieties, and on Adele groups are given.

1.4 ERGODIC THEOREMS FOR LATTICE SUBGROUPS

We now turn to our third main result, namely, the solution to the problem of establishing ergodic theorems for a general action of a lattice subgroup on a probability space (X, μ). This result uses Theorem 1.2 as a basic tool; namely, it is applied to the action of G induced by the action of Γ on (X, μ). The argument generalizes the one used in the proof of Theorem 1.5, where we consider the action of G induced from the trivial action of Γ on a point. However, the increased generality requires considerable further effort and additional arguments.

To formulate the result, consider the set of lattice points $\Gamma_t = \Gamma \cap G_t$. Let λ_t denote the probability measure on Γ uniformly distributed on Γ_t.

We begin with the following fundamental mean ergodic theorem for arbitrary lattice actions.

Theorem 1.7. Mean ergodic theorem for lattice actions. *Let G, G_t, and Γ be as in Theorem 1.5. Let (X, μ) be an ergodic measure-preserving action of Γ.*

1. *Assume that the action of G induced from the Γ-action on (X, μ) is irreducible or that the sets G_t are balanced. Then for every $f \in L^p(X)$, $1 \le p < \infty$,*

$$\lim_{t \to \infty} \left\| \frac{1}{|\Gamma_t|} \sum_{\gamma \in \Gamma_t} f(\gamma^{-1}x) - \int_X f d\mu \right\|_{L^p(X)} = 0.$$

2. *Assume that the action of G induced from the Γ-action on (X, μ) has a strong spectral gap or that it has a spectral gap and the sets G_t are well balanced. Then for every $f \in L^p(X)$, $1 < p < \infty$,*

$$\left\| \frac{1}{|\Gamma_t|} \sum_{\gamma \in \Gamma_t} f(\gamma^{-1}x) - \int_X f d\mu \right\|_{L^p(X)} \le C e^{-\delta_p t} \|f\|_{L^p(X)},$$

where $\delta_p > 0$ is determined explicitly by the spectral gap for the induced G-action (and also depends on the family G_t).

One immediate application of Theorem 1.7 arises when we take X to be a transitive action on a finite space, namely, $X = \Gamma/\Delta$, Δ a finite index subgroup.

Corollary 1.8. Equidistribution in finite actions. *Let G, Γ, and G_t be as in Theorem 1.5. Let $\Delta \subset \Gamma$ be a subgroup of finite index and γ_0 any element in Γ. Under the assumptions of Theorem 1.7(2),*

$$\frac{1}{|\Gamma_t|} \cdot |\{\gamma \in \Gamma \cap G_t : \gamma \cong \gamma_0 \bmod \Delta\}| = \frac{1}{[\Gamma : \Delta]} + O(e^{-\delta t}),$$

where $\delta > 0$ and is determined explicitly by the spectral gap in G/Δ.

We remark that a weaker conclusion than Corollary 1.8, namely, equidistribution of the lattice points in $\Gamma \cap G_t$ among the cosets of Δ in Γ, can also be obtained using the method in [GW], which employs Ratner's theory of unipotent flow. It is also possible to derive the main term from considerations related to the mixing property of flows on G/Γ.

Another application of the mean ergodic theorem is in the proof of an equidistribution theorem for the corresponding averages in isometric actions of the lattice. The result is as follows.

Theorem 1.9. Equidistribution in isometric actions of lattices. *Let G, G_t, and Γ be as in Theorem 1.7. Let (S, d) be a compact metric space on which Γ acts by isometries and assume that the action is ergodic with respect to an invariant probability measure μ whose support coincides with S. Then under the assumptions of Theorem 1.7(1), for every continuous function f on S and every point $s \in S$,*

$$\lim_{t \to \infty} \frac{1}{|\Gamma_t|} \sum_{\gamma \in \Gamma_t} f(\gamma^{-1}s) = \int_S f \, d\mu,$$

and the convergence is uniform in $s \in S$ (i.e., in the supremum norm on $C(S)$).

Let us now formulate pointwise ergodic theorems for general actions of lattices.

Theorem 1.10. Pointwise ergodic theorems for general lattice actions. *Let G, G_t, Γ, and (X, μ) be as in Theorem 1.7.*

1. *Assume that the action induced to G is irreducible and the averages β_t are left-radial. Then the averages λ_t satisfy the pointwise ergodic theorem in $L^p(X)$, $1 < p < \infty$; namely, for $f \in L^p(X)$ and almost every $x \in X$,*

$$\lim_{t \to \infty} \frac{1}{|\Gamma_t|} \sum_{\gamma \in \Gamma_t} f(\gamma^{-1}x) = \int_X f \, d\mu.$$

The same conclusion also holds when the induced action is reducible, provided β_t are standard, well balanced, and boundary-regular.

2. *Retain the assumption of Theorem 1.7(2). Then the convergence of λ_t to the ergodic mean is almost surely exponentially fast; namely, for $f \in L^p(X)$, $1 < p < \infty$, and almost every $x \in X$,*

$$\left| \frac{1}{|\Gamma_t|} \sum_{\gamma \in \Gamma_t} f(\gamma^{-1}x) - \int_X f \, d\mu \right| \leq C_p(x, f) e^{-\zeta_p t},$$

where $\zeta_p > 0$ is determined explicitly by the spectral gap for the induced G-action (and the family G_t).

Remark 1.11.

1. Note that if G is simple, then of course any action of G induced from an ergodic action of a lattice subgroup is irreducible. However, if G is not simple, then the induced action can be reducible and then the assumption that the averages are balanced is necessary in Theorem 1.10(1). We assume in fact that they are standard, well balanced, and boundary-regular, as we will apply Theorem 1.2(1) to the induced action.

2. Note further that if G is simple and has property T, then the assumption of a strong spectral gap stated in Theorem 1.10(2) is satisfied for every ergodic action of every lattice subgroup. Furthermore, in that case ζ_p has an explicit positive lower bound depending on G and G_t only and independent of Γ and X.

3. It may be the case that whenever G/Γ has a strong spectral gap, so does every action of G induced from an ergodic action of the irreducible lattice Γ which has a spectral gap, but this problem also seems to be open.

4. As we shall see in §6.1, the possibility of utilizing the induced G-action to deduce information on *pointwise convergence* in the inducing Γ-action depends on the invariance principle stated in Theorem 1.3 for admissible averages on G.

Further, below we will give a complete analysis valid for S-algebraic groups and their lattices in all cases, but let us here demonstrate our results in a more concrete fashion which shows, in particular, that sets Γ_t satisfying all the assumptions required do exist. Indeed, let G be a connected semisimple Lie group with finite center and no nontrivial compact factors. Let G/K be its symmetric space, let d be the Riemannian distance associated with the Cartan-Killing form, and let β_t be the Haar-uniform averages on the sets

$$G_t = \{g \in G \,;\, d(gK, K) \le t\}. \tag{1.3}$$

Then G_t are admissible and well balanced, and it has been established in [N2], [N3], [NS2], and [MNS] that in every ergodic probability measure–preserving action of G, the family β_t satisfies the pointwise ergodic theorem in L^p, $1 < p < \infty$, as in Theorem 1.2(1). Furthermore, if the action has a spectral gap, then the convergence to the ergodic mean is exponentially fast, as in Theorem 1.2(2).

Now let $\Gamma \subset G$ be any lattice subgroup. Then the following result, announced in [N6, Thm. 14.4], holds.

Theorem 1.12. Ergodic theorems for lattice points in Riemannian balls. *Let G, G_t, and Γ be as in the preceding paragraph and λ_t the uniform averages on $\Gamma \cap G_t$. Then in every probability measure–preserving action of Γ, λ_t satisfy the mean ergodic theorem in L^p, $1 \le p < \infty$, and the pointwise ergodic theorem in L^p, $1 < p \le \infty$. If the Γ-action has a spectral gap, then λ_t satisfy the exponential mean and pointwise ergodic theorem as in Theorems 1.7(2) and 1.10(2). Finally, λ_t satisfy the equidistribution theorem w.r.t. an ergodic invariant probability measure of full support in every isometric action of Γ.*

1.5 SCOPE OF THE METHOD

Motivated by concrete lattice point–counting problems, by equidistribution problems, and by other applications of ergodic theorems, we have attempted to give a unified and comprehensive treatment for a large class of averages on all S-algebraic groups and their lattice subgroups, which applies to every ergodic action of the group or the lattice. This level of generality inevitably brings up several issues that must be addressed, including the following:

1. Natural examples such as $SL_2(\mathbb{Z}[\sqrt{2}])$, which is an irreducible lattice in the group $SL_2(\mathbb{R}) \times SL_2(\mathbb{R})$, necessitate the consideration of general semisimple Lie groups.

2. Natural examples such as $SL_2(\mathbb{Z}[\frac{1}{p}])$, which is an irreducible lattice in the group $SL_2(\mathbb{R}) \times SL_2(\mathbb{Q}_p)$, necessitate the consideration of products of algebraic groups over different fields.

3. Natural lattice point–counting problems, such as the problem of integral equivalence of homogeneous forms that we will discuss below, require the consideration of a wide variety of distances, going beyond norms or invariant distances on symmetric spaces and beyond radial averages.

4. Consideration of all ergodic actions of product groups necessitates the analysis of reducible actions, namely, actions where some component of the product group does not act ergodically. This raises the issue of whether the mass distribution of the balls G_t among the simple factors is balanced, or well balanced, in a sense to be made precise below.

5. Consideration of product groups necessitates the analysis of actions with a spectral gap, but where some component group acts without a spectral gap, namely, with an asymptotically invariant sequence.

6. The fact that totally disconnected linear algebraic groups such as $PGL_2(\mathbb{Q}_p)$ can admit finite-dimensional irreducible nontrivial permutation representations where the ergodic theorems fail necessitates restricting our attention to actions where each component of the group is mixing in the orthogonal complement of the space of its invariants.

We note that resolution of all the issues listed above is essential in the proof of the ergodic theorems for lattice subgroups. Indeed, the basic underlying principle of our method of proof is to induce an action of Γ to an action of G and reduce the ergodic theorems for λ_t to those of β_t. However, it may perhaps be the case that the resulting action of G is reducible, or that it may perhaps have a spectral gap but not a strong spectral gap. In these cases, whether the averages are balanced or well balanced becomes a necessary consideration. We will introduce the tools necessary for a systematic development of the general theory in Chapter 3 and employ them in our subsequent proofs of the ergodic theorems.

As is already clear from the statements of the results above, the distinction between actions with and without a spectral gap is fundamental in determining which ergodic theorems apply, and the two cases call for rather different methods of proof. Thus the results will be established according to the following scheme:

1. Ergodic theorems for averages on semisimple S-algebraic groups in the presence of a spectral gap.

2. Ergodic theorems for averages on semisimple S-algebraic groups in the absence of a spectral gap.

3. Stability of admissible averages on semisimple S-algebraic groups and an invariance principle for their ergodic actions.

4. Mean, maximal, and pointwise ergodic theorems for lattice subgroups in the absence of a spectral gap.

5. Exponential pointwise ergodic theorem for lattice actions in the presence of a spectral gap.

6. Equidistribution for isometric lattice actions.

As we shall see below, this scheme applies in a much wider context than that of semisimple S-algebraic groups. We will formulate in Chapter 6 a general recipe to derive ergodic theorems for actions of a lattice subgroup Γ, given that the underlying averages on the enveloping lcsc group G satisfy the corresponding ergodic theorems. We will also demonstrate in Chapter 5 that the ergodic theorems for G-actions hold, provided only that certain natural spectral, geometric, and regularity conditions are satisfied by the group G and the sets G_t.

Chapter Two

Examples and applications

Let us now consider some examples and applications of the results stated in Chapter 1 and compare our results to some precedents in the literature.

2.1 HYPERBOLIC LATTICE POINTS PROBLEM

We begin by applying Theorem 1.5 to the classical lattice point–counting problem in hyperbolic space. Let us call a lattice subgroup Γ tempered if the spectrum of the representation of the isometry group in $L_0^2(G/\Gamma)$ is *tempered*, namely, the representation is weakly contained in the regular representation of the isometry group, or equivalently (see [CHH]) for a dense family of functions, the corresponding matrix coefficients are in $L^{2+\varepsilon}(G)$ for every $\varepsilon > 0$.

Corollary 2.1. *Let $\mathbb{H}^n$ be hyperbolic n-space taken with constant curvature -1 and the resulting volume form. Let B_t be the Riemannian balls centered at a given point and let Γ be any lattice. Then for all $\varepsilon > 0$,*

1. Provided that $\left\| \pi_{G/\Gamma}(\beta_t) \right\|_{L_0^2(G/\Gamma)} = O_\varepsilon(e^{-t(\theta-\varepsilon)})$, we have

$$\frac{|\Gamma \cap B_t|}{\mathrm{vol}(B_t)} = \frac{1}{\mathrm{vol}(\mathbb{H}^n/\Gamma)} + O_\varepsilon\left(\exp -t\left(\frac{\theta}{n+1} - \varepsilon\right)\right).$$

2. In particular, if Γ is tempered, then

$$\frac{|\Gamma \cap B_t|}{\mathrm{vol}(B_t)} = \frac{1}{\mathrm{vol}(\mathbb{H}^n/\Gamma)} + O_\varepsilon\left(\exp -t\left(\frac{n-1}{2(n+1)} - \varepsilon\right)\right).$$

We remark that the bound stated above is actually better than that provided by Theorem 1.5, as the error term here is given by $\theta/(\dim G/K + 1)$ rather than $\theta/(\dim G + 1)$. This is a consequence of the fact that here we have taken bi-K-invariant averages on G, so that the arguments used in the proof of Theorem 1.5 can be applied on $K \backslash G$ rather than G. The same bound holds for any choice of bi-K-invariant admissible sets B_t. The spectral gap parameter is given by $\theta = \frac{1}{2}(n-1)$ in the tempered case since the convolution norm of β_t on $L^2(G)$ is dominated by $\mathrm{vol}(B_t)^{-1/2+\varepsilon}$ (see §5.2.2.2), which is asymptotic to $\exp -t\left(\frac{1}{2}(n-1) - \varepsilon\right)$ (recall that vol B_t is asymptotic to $c_n e^{(n-1)t}$).

For comparison, the best existing bound for a tempered lattice in hyperbolic n-space ($n \geq 2$) is due to Selberg [Se] and Lax and Phillips [LP] and is given by

$$\frac{|\Gamma \cap B_t|}{\mathrm{vol}\, B_t} = \frac{1}{\mathrm{vol}(\mathbb{H}^n/\Gamma)} + O_\varepsilon\left(\exp -t\left(\frac{n-1}{n+1} - \varepsilon\right)\right).$$

The method developed in [LP] uses detailed estimates on solutions to the wave equation, and in [Se] the method uses refined properties of the spectral expansion associated with the Harish-Chandra spherical transform. In particular, both methods require that B_t be bi-K-invariant sets.

On the other hand, the estimate of Theorem 1.5 holds for *any* family of admissible sets B_t. Thus even the following sample corollary is new for $G = \mathrm{PSL}_2(\mathbb{R})$. Define for $1 \le r < \infty$, $\|A\|_r = \left(\sum_{i,j=1}^2 |a_{i,j}|^r \right)^{1/r}$, and $\|A\|_\infty = \max |a_{i,j}|$.

Corollary 2.2. *For any tempered finite-covolume Fuchsian group and for any $1 \le r \le \infty$, with the normalization $\mathrm{vol}(G/\Gamma) = 1$,*

$$\frac{|\{\gamma \in \Gamma \,;\, \|\gamma\|_r \le T\}|}{\mathrm{vol}\{g \in \mathrm{SL}_2(\mathbb{R})\,;\, \|g\|_r \le T\}} = 1 + O_{\varepsilon,r}\left(T^{-1/4}\right).$$

In particular, this holds for $\Gamma = \mathrm{PSL}_2(\mathbb{Z})$.

By Theorem 1.5, an explicit estimate for the error term in the lattice point–counting problem in admissible domains in hyperbolic space of arbitrary dimension depends only on the rate of decay of $\|\pi^0_{G/\Gamma}(\beta_t)\|$. This quantity in turn has a lower bound which depends only on the rate of volume growth of B_t and on a lower bound for the spectrum of the Laplacian on $K\backslash G/\Gamma$, see §8.1.

2.2 COUNTING INTEGRAL UNIMODULAR MATRICES

Let $G = \mathrm{SL}_n(\mathbb{R})$, $n \ge 2$, be the group of unimodular matrices and $\Gamma = \mathrm{SL}_n(\mathbb{Z})$ the subgroup of integral matrices. A natural choice of balls here are those defined by taking the defining representation and the rotation-invariant linear norm on $\mathrm{M}_n(\mathbb{R})$ given by $(\mathrm{tr}\, A^t A)^{1/2}$. Let B'_T denote the norm ball of radius T intersected with $\mathrm{SL}_n(\mathbb{R})$. Here the best result to date is due to [DRS] and is given by

$$\frac{|\Gamma \cap B'_T|}{\mathrm{vol}\, B'_T} = 1 + O_\varepsilon\left(T^{-1/(n+1)+\varepsilon}\right).$$

Letting $t = \log T$, the family $B_t = B'_{e^t}$ is admissible. For our estimate, we need to bound θ, the essential rate of decay of $\|\pi_{G/\Gamma}(\beta_t)\|$ in $L^2_0(G/\Gamma)$. For $n = 2$, $\theta = 1/2$ as noted above since the representation is tempered. For $n \ge 3$, $\mathrm{SL}_n(\mathbb{R})$ has property T, and we can simply use a bound valid for all of its representations simultaneously (provided only that they contain no invariant unit vectors). Note that in the case of $L^2_0(\mathrm{SL}_n(\mathbb{R})/\mathrm{SL}_n(\mathbb{Z}))$ this also happens to be the best possible estimate since the spherical function with slowest decay does in fact occur in the spectrum. According to [DRS], every nonconstant spherical function on $\mathrm{SL}_n(\mathbb{R})$ is in L^p for $p > p(\mathrm{SL}_n(\mathbb{R})/\mathrm{SL}_n(\mathbb{Z})) = 2(n-1)$. This implies (see [CHH] and Theorem 5.4 below) that the matrix coefficients of π have an estimate in terms of $\Xi_G^{1/(n-1)+\varepsilon}$, where Ξ_G is the Harish-Chandra function. Using the standard estimate for Ξ_G (see, for instance, [GV]) and Corollary 8.1(3),

$$\|\pi(\beta_t)\| \le \left(C_0 \,\mathrm{vol}(B_t)^{-1/2+\varepsilon}\right)^{1/(n-1)} \le C \exp\left(-t\left(\frac{n^2-n}{2(n-1)} - \varepsilon\right)\right),$$

where the last estimate uses the fact that (see [DRS])

$$\text{vol}(B_T') = \text{vol}(\{g \in \text{SL}_n(\mathbb{R}) ; \|g\|_2 \leq T\}) \cong c_n T^{n^2-n}.$$

Therefore we have the estimate $\theta = n/2$, so that $\theta/(\dim G + 1) = 1/(2n)$ and $\theta/(\dim G/K + 1) = 1/(n+1)$. Thus we recapture the bound given by [DRS] for the case of balls defined by a rotation-invariant norm.

More generally, letting n_e denote the least even integer such that

$$n_e \geq p(\text{SL}_n(\mathbb{R})/\text{SL}_n(\mathbb{Z}))/2 = n - 1,$$

we have the following corollary by Theorem 1.5 and §5.2.2.2.

Corollary 2.3. *For any family of admissible sets $B_t \subset \text{SL}_n(\mathbb{R})$, in particular those defined by any norm on $M_n(\mathbb{R})$, and for any lattice subgroup Γ, the following bound holds:*

$$\frac{|\Gamma \cap B_t|}{\text{vol } B_t} = 1 + O_\varepsilon \left(\text{vol}(B_t)^{-1/(2n^2 n_e)+\varepsilon}\right).$$

We note that the method in [DRS] utilizes the commutativity of the algebra of bi-K-invariant L^1-functions on G. Extending this method beyond the case of bi-K-invariant sets is possible in principle but would require further elaboration regarding the spectral analysis of K-finite functions, decreasing the quality of the error term.

Recently, F. Maucourant [Ma] has obtained a bound for the lattice point–counting problem for certain simple groups and certain norms, subject to some constraints. Thus for the standard representation of $\text{SL}_n(\mathbb{R})$, when $n \geq 7$ the error estimate obtained in [Ma] is $1/(6n)$, which is weaker than the estimate of $1/(2n)$ above. That is also the case for $3 \leq n \leq 6$. The case $n = 2$ is not addressed in [Ma].

2.3 INTEGRAL EQUIVALENCE OF GENERAL FORMS

2.3.1 Binary forms

Let us revisit the problem of integral equivalence of binary forms considered in [DRS]. Let W_k denote the vector space of homogeneous binary forms of degree $k \geq 3$:

$$W_k = \left\{ f(x, y) = a_0 x^k + a_1 x^{k-1} y + \cdots + a_k y^k : a_0, \ldots, a_k \in \mathbb{R} \right\}.$$

The group $\text{SL}_2(\mathbb{R})$ acts on $W_k(\mathbb{R})$ by acting linearly on the variables of the form, and when $k \geq 3$, the stability group of a generic form is finite. Two forms are in the same $\text{SL}_2(\mathbb{R})$-orbit if and only if they are equivalent under a linear substitution, and two forms are in the same $\text{SL}_2(\mathbb{Z})$-orbit iff they are integrally equivalent. Fix *any* norm on $W_k(\mathbb{R})$, one example being the norm considered in [DRS]:

$$\|f\|^2 = \|(a_0, \ldots, a_k)\|^2 = \sum_{i=0}^{k} \binom{k}{i}^{-1} a_i^2.$$

The orbits of $\text{SL}_2(\mathbb{R})$ are closed, and for each orbit we can consider the lattice point–counting problem, or equivalently, the problem of counting forms integrally equivalent to a given form. Thus fix some f_0 with a finite stabilizer and nonzero discriminant and denote $B'_T = \{f ; f \cong_{\mathbb{R}} f_0, \|f\| \leq T\}$. We note that it has been established in [DRS] that $\mathrm{vol}(B'_T) \sim cT^{2/k}$. We further assume that the form satisfies $f_0(x, y) \neq 0$ for $(x, y) \neq (0, 0)$. We then have the following corollary of Theorem 1.5.

Corollary 2.4. *Notation being as above, the number of forms integrally equivalent to f_0 of norm at most T is estimated by*

$$\left| \frac{|\{f ; f \cong_{\mathbb{Z}} f_0, \|f\| \leq T\}|}{\mathrm{vol}(B'_T)} - \frac{1}{\left| St_{SL_2(\mathbb{Z})}(f_0) \right|} \right|$$

$$\leq C(\varepsilon, k, f_0) \, \mathrm{vol}(B'_T)^{-1/8+\varepsilon} \leq C'(\varepsilon, k, f_0) T^{-1/(4k)+\varepsilon} .$$

Indeed, the problem under consideration is simply that of counting the number of $\gamma \in SL_2(\mathbb{Z})$ satisfying $\|\tau_k(\gamma) f_0\| \leq T$ for a particular choice of finite-dimensional representation τ_k of $SL_2(\mathbb{R})$ and a particular choice of distance function on G, namely, the restriction of a norm on the representation space to the G-orbit of f_0. Thus the corollary is an immediate consequence of the fact that the sets $B_t = B'_{e^t}$ are admissible (see §7.4) together with Corollary 2.3 and the fact that the representation of $SL_2(\mathbb{R})$ on $SL_2(\mathbb{R})/SL_2(\mathbb{Z})$ is tempered. Note that here the sets B_t are definitely not radial.

We note that the existence of the limit was established in [DRS, Thm. 1.9]. The method of proof employed there can in principle be made effective and produce an error estimate.

2.3.2 Integral equivalence of forms in many variables

Our considerations are not limited to binary forms, and we can consider the problem of integral equivalence, as well as simultaneous integral equivalence, of forms in any number of variables. Thus let $W_{n,k}$ be the real vector space of forms of degree k in n variables. The group $SL_n(\mathbb{R})$ admits a representation $\sigma_{n,k}$ on $W_{n,k}$ by acting linearly on the variables. Fix any norm on $W_{n,k}$. As before, n_e denotes the least even integer such that $n_e \geq p(SL_n(\mathbb{R})/SL_n(\mathbb{Z}))/2 = n - 1$.

Consider a form f_0 with a compact stability group. Let us assume that $f_0(x) \neq 0$ for $x \neq 0$, so that the projection of the vectors $u f_0$ onto the highest weight subspace of $W_{n,k}$ never vanishes as u ranges over the associated maximal compact subgroup. Let C'_T denote the set of forms equivalent to f_0 and of norm at most T. Then $C_t = C'_{e^t}$ is an admissible family (see §7.4, where a volume asymptotic is also established). Hence Corollary 2.3 applies and yields the following.

Corollary 2.5. Integral equivalence of forms in many variables. *Notation and assumptions, being as in the preceding paragraph, we have*

$$\left| \frac{|\{f ; f \cong_{\mathbb{Z}} f_0, \|f\| \leq T\}|}{\mathrm{vol}(C'_T)} - \frac{1}{\left| St_{SL_n(\mathbb{Z})}(f_0) \right|} \right|$$

$$\leq C(\varepsilon, n, k, f_0) \, \mathrm{vol}(C'_T)^{-1/(2n^2 n_e)+\varepsilon} .$$

Now let $f_1, \ldots, f_N$ be a fixed (but arbitrary) ordered basis of $W_{n,k}$ ($N = \dim W_{n,k}$). We can consider ordered bases $f'_1, \ldots, f'_N$ which are integrally equivalent to it, namely, $f_i \cong_{\mathbb{Z}} f'_i$, $1 \leq i \leq N$. Let

$$D'_T = \left\{ g \in \mathrm{SL}_k(\mathbb{R}) \, ; \, \| g f'_i \| \leq T \, , \, 1 \leq i \leq N \right\}.$$

Then $D_t = D'_{e^t}$ is in fact a family of balls defined by a norm in a linear representation, and any such family is admissible (see Chapter 7). Here the stability group is necessarily trivial, so again from Corollary 2.3 we have the following.

Corollary 2.6. Simultaneous integral equivalence. *Notation and assumptions being as in the preceding paragraph, we have*

$$\left| \frac{\left| \left\{ (f'_1, \ldots, f'_N) \, ; \, f'_i \cong_{\mathbb{Z}} f_i \, , \, \| f'_i \| \leq T \, , \, 1 \leq i \leq N \right\} \right|}{\mathrm{vol}(D'_T)} - 1 \right|$$

$$\leq C(\varepsilon, n, k, f_i, \ldots, f_N) \, \mathrm{vol}(D'_T)^{-1/(2n^2 n_e)+\varepsilon}.$$

2.4 LATTICE POINTS IN S-ALGEBRAIC GROUPS

As already noted, all of our results will in fact be formulated and proved in the context of semisimple S-algebraic groups. Let us demonstrate them in the following basic case as a motivation for the developments below.

Let p be a prime and consider $G_n = \mathrm{PSL}_n(\mathbb{R}) \times \mathrm{PSL}_n(\mathbb{Q}_p)$ and the S-arithmetic lattice $\Gamma_n = \mathrm{PSL}_n(\mathbb{Z}[\frac{1}{p}])$. Take the norm on $M_n(\mathbb{R})$ whose square is $\mathrm{tr}\, A^t A$ and its (well-defined) restriction to $\mathrm{PSL}_n(\mathbb{R})$. For $A \in \mathrm{PSL}_n(\mathbb{Q}_p)$ let $|A|_p = \max_{1 \leq i, j \leq n} |a_{i,j}|_p$, where $|a|_p$ is the p-adic absolute value of $a \in \mathbb{Q}_p$, normalized as usual by $|p|_p = \frac{1}{p}$. If $A \in M_n(\mathbb{Z})$, we write $(A, p) = 1$ if $(a_{i,j}, p) = 1$ for some entry $a_{i,j}$. Define the height function on G_n by $H(A, B) = \| A \| \, |B|_p$.

Let C_T be the set of integral matrices with Euclidean norm bounded by T and with $\det A$ a power of p^n and $(A, p) = 1$, namely,

$$C_T = \left\{ A \in M_n(\mathbb{Z}) \, ; \, \mathrm{tr}\, A^t A \leq T^2, \det A \in p^{n\mathbb{N}}, (A, p) = 1 \right\}.$$

Proposition 2.7. *The family C_T satisfies*

$$\frac{|C_T|}{\mathrm{vol}\, \{ g \in G_n \, ; \, H(g) \leq T \}} = 1 + O_{\varepsilon, p, n} \left(T^{-1/2n+\varepsilon} \right).$$

The proposition is a consequence of Corollary 8.1 and the fact that the set C_T in question is in one-to-one correspondence with a set of lattice points in balls B_t ($t = \log T$) in G_n defined by the natural height function. Indeed, for $y = (u, v) \in G_n$ the height is $H(y) = \sqrt{\mathrm{tr}\, u^t u} \cdot |v|_p$. Clearly, if $u \in \mathrm{PSL}_n(\mathbb{Z}[\frac{1}{p}])$ and $|u|_p = p^k$ (where $k \geq 0$), then $A = p^k u \in M_n(\mathbb{Z})$, $(A, p) = 1$, and $\det A = p^{kn} \det u \in p^{n\mathbb{N}}$. Also, $\| A \| = \| p^k u \| = p^k \| u \| = H(\gamma)$, where $\gamma = (u, u) \in \Gamma$, so that C_T maps bijectively with $\{ y \in \Gamma \, ; \, H(\gamma) \leq T \} = \Gamma \cap B'_T$, where $B'_T = \{ y \in G \, ; \, H(y) \leq T \}$.

Now consider the basis of open sets at the identity in G_n given by $\mathcal{O}_\varepsilon = \mathcal{U}_\varepsilon \times \mathcal{K}_p$, the product of Riemannian balls $\mathcal{U}_\varepsilon$ on $\mathrm{PSL}_n(\mathbb{R})$ and the compact open neighborhood:

$$\mathcal{K}_p = \left\{ v \in \mathrm{PSL}_n(\mathbb{Q}_p) ;\ |v - I|_p \leq 1 \right\}.$$

Defining $B_t = B'_{e^t}$, the family B_t is admissible w.r.t. to $\mathcal{O}_\varepsilon$. This follows from Theorem 3.15(4) since the height is defined by a product of two norms. The unitary representation of G_n on $L_0^2(G_n/\Gamma_n)$ is strongly $L^{2(n-1)+\varepsilon}$, and hence, since β_t are radial, we deduce from Corollary 8.1(3) that

$$\|\pi_0(\beta_t)\| \leq \mathrm{vol}(B_t)^{-1/(2(n-1))+\varepsilon}.$$

A direct calculation of the volume of B_T shows that $\mathrm{vol}(B'_T) \geq C_\varepsilon T^{n^2-n+\varepsilon}$, and this gives the error term above.

2.5 EXAMPLES OF ERGODIC THEOREMS FOR LATTICE ACTIONS

2.5.1 Exponentially fast convergence on the torus

Fix a norm on $\mathrm{M}_n(\mathbb{R})$ and consider the corresponding norm balls $G_t \subset \mathrm{SL}_n(\mathbb{R})$ and the averages λ_t on $\Gamma_t = \mathrm{SL}_n(\mathbb{Z}) \cap G_t$.

The following result is a direct corollary of Theorem 1.10 and the well-known fact that the action of $\mathrm{SL}_n(\mathbb{Z})$ on the n-torus $\mathbb{T}^n$ admits a spectral gap.

Corollary 2.8. *Consider the action of $\mathrm{SL}_n(\mathbb{Z})$ on $(\mathbb{T}^n, m)$, where m is the normalized Lebesgue measure. The averages λ_t satisfy for every $f \in L^p(X), 1 < p < \infty$, for almost every $x \in X$,*

$$\left| \lambda_t f(x) - \int_{\mathbb{T}^n} f\, dm \right| \leq C_p(f, x) e^{-\eta_n t},$$

where $\eta_n > 0$ is explicit.

2.5.2 Exponentially fast convergence in the space of unimodular lattices

Let Γ be a lattice in a simple noncompact Lie group H and $\tau : H \to \mathrm{SL}_n(\mathbb{R})$ a rational representation with finite kernel. Fix a norm on $\mathrm{M}_n(\mathbb{R})$ and let $H_t = \{h \in H ;\ \log \|\tau(h)\| < t\}$. Then the uniform averages λ_t on $\Gamma \cap H_t$ converge exponentially fast to the ergodic mean in any of the actions of Γ of $\mathrm{SL}_n(\mathbb{R})/\Delta$, Δ a lattice subgroup. In particular, letting $\Delta = \mathrm{SL}_n(\mathbb{Z})$, the homogeneous space $\mathcal{L}_n = \mathrm{SL}_n(\mathbb{R})/\mathrm{SL}_n(\mathbb{Z})$ can be identified with the space of unimodular lattices in $\mathbb{R}^n$. For such a lattice $L \in \mathcal{L}_n$, let $f(L)$ be the number of vectors in L whose length (w.r.t. the standard Euclidean norm) is at most 1. Then for $n \geq 2$, $f \in L^p(\mathcal{L}_n)$, $1 \leq p < n$, and we let $\kappa_n = \int_{\mathcal{L}_n} f(L)\, dm(L)$ denote the average number of vectors of length at most 1 in a unimodular lattice L. Note that by Siegel's formula κ_n equals the volume of the unit ball in $\mathbb{R}^n$.

We can now appeal to Theorems 1.7 and 1.10 and apply them to the averages λ_t^H. We conclude the following.

Corollary 2.9. *Let $n \geq 2$ and $1 < p < n$. Then for almost every unimodular lattice $L \in \mathcal{L}_n$, we have*

$$\frac{\#\{\gamma \in \Gamma \cap H_t ; \; |\#(\gamma L \cap B_1(0)) - \kappa_n| \geq \delta\}}{\#\{\gamma \in \Gamma \cap H_t\}} \leq C_p \delta^{-p} \|f\|^p_{L^p(\mathcal{L}_n)} e^{-\zeta_{p,n}t},$$

where $\zeta_{p,n} > 0$ is explicit and depends on the spectral gap of the H-action on $L^2(\mathcal{L}_n)$ and the admissible family H_t.

2.5.3 Equidistribution and exponentially fast convergence

Let us now consider the case where the lattice Γ acts isometrically on a compact metric space, preserving a ergodic probability measure of full support. Two important families of examples are given by the following.

1. The action of Γ on any of its profinite completions, with the invariant probability measure being the Haar measure on the compact group. In particular, this includes the congruence completion when Γ is arithmetic.

2. The action of Γ on the unit sphere in $\mathbb{C}^n$ or $\mathbb{R}^n$ via a finite-dimensional unitary or orthogonal representation with a dense orbit on the unit sphere (when such exists).

We note that when combining Theorems 1.9 and 1.10, the following interesting phenomenon emerges.

Corollary 2.10. *Let Γ be a lattice subgroup in a connected almost simple non-compact Lie group with property T. Let G_t be admissible and λ_t the averages uniformly distributed on $G_t \cap \Gamma$. Then in every isometric action of Γ on a compact metric space S, ergodic with respect to a probability measure m of full support, the following holds. For every continuous function $f \in C(S)$, $\lambda_t f(s)$ converges to $\int_S f \, dm$ for every $s \in S$ and converges exponentially fast to $\int_S f \, dm$ for almost every $s \in S$. The exponential rate of convergence depends only on G_t and G and is independent of S and Γ.*

2.5.4 Ergodic theorems for free groups

Let us note some further ergodic theorems which follow from Theorem 1.10.

1. The index 6 principal level 2 congruence group $\Gamma(2)$ of $\mathrm{PSL}_2(\mathbb{Z})$ is a free group on two generators. Theorem 1.10 thus gives new ergodic theorems for *arbitrary actions* of free groups, where the averages taken are uniformly distributed on, say, norm balls. If the free group action has a spectral gap, the convergence is exponentially fast. These averages are completely different from the averages w.r.t. a word metric on the free group discussed in [N1] and [NS1].

2. Note that for the averages just described, the phenomenon of periodicity (see [N6, §10.5]) associated with the existence of the sign character of the free group does not arise: the limit is always the ergodic mean.

Thus, in particular, Theorem 1.8 implies that for any norm on $M_2(\mathbb{R})$, norm balls become equidistributed among the cosets of any finite index subgroup of $\Gamma(2) \cong \mathbb{F}_2$ at an exponentially fast rate.

3. Similar comments also apply, for example, to the lattice $\Gamma = \mathrm{PSL}_2(\mathbb{Z}) = \mathbb{Z}_2 * \mathbb{Z}_3 \subset \mathrm{PSL}_2(\mathbb{R})$ itself, and again the averages in question are different from the word metric ones discussed in [N1]. Another family of examples are lattices in $\mathrm{PGL}_3(\mathbb{Q}_p)$, to which our results stated in §1.4 apply. In particular, this includes the lattices acting simply transitively on the vertices of the Bruhat-Tits building, generalizing [N6, Thm. 11.10] for these lattices.

Chapter Three

Definitions, preliminaries, and basic tools

In this chapter we introduce the necessary tools and definitions which will allow us to develop ergodic theorems on general locally compact second countable groups in a systematic fashion. We refer the reader to §1.5, where the motivation for some of the concepts appearing below is explained.

3.1 MAXIMAL AND EXPONENTIAL-MAXIMAL INEQUALITIES

Let G be an lcsc group with a left-invariant Haar measure m_G. Let $(X, \mathcal{B}, \mu)$ be a standard Borel space with a Borel-measurable G-action preserving the probability measure μ. There is a natural isometric representation π_X of G on the spaces $L^p(\mu)$, $1 \leq p \leq \infty$, defined by

$$(\pi_X(g)f)(x) = f(g^{-1}x), \quad g \in G, \ f \in L^p(\mu).$$

To each finite Borel measure β on G, we associate the bounded linear operator

$$(\pi_X(\beta)f)(x) = \int_G f(g^{-1}x) \, d\beta(g)$$

acting on $L^p(\mu)$. In particular, given an increasing sequence G_t, $t > 0$, of Borel subsets of positive finite measure of G, we consider the Borel probability measures given by the Haar-uniform averages over the sets G_t:

$$\beta_t = \frac{1}{m_G(G_t)} \int_{G_t} \delta_g \, dm_G(g) \tag{3.1}$$

and the associated operators $\pi_X(\beta_t)$.

Definition 3.1. Maximal inequalities and ergodic theorems. Let ν_t, $t > 0$, be a one-parameter family of absolutely continuous probability measures on G such that the map $t \mapsto \nu_t$ is continuous in the $L^1(G)$-norm. The maximal function $\sup_{t > t_0} |\pi_X(\nu_t)f|$, $f \in L^\infty(X)$, is then measurable. We define the following.

1. The family ν_t satisfies the *strong maximal inequality in* $(L^p(\mu), L^r(\mu))$, $p \geq r$, if there exist $t_0 \geq 0$ and $C_{p,r} > 0$ such that for every $f \in L^p(\mu)$,

$$\left\| \sup_{t > t_0} |\pi_X(\nu_t)f| \right\|_{L^r(\mu)} \leq C_{p,r} \|f\|_{L^p(\mu)}.$$

2. The family ν_t satisfies the *mean ergodic theorem in* $L^p(\mu)$ if for every $f \in L^p(\mu)$,

$$\left\| \pi_X(\nu_t)f - \int_X f \, d\mu \right\|_{L^p(\mu)} \to 0 \quad \text{as } t \to \infty.$$

3. The family v_t satisfies the *pointwise ergodic theorem in $L^p(\mu)$* if for every $f \in L^p(\mu)$,

$$\pi_X(v_t)f(x) \to \int_X f \, d\mu \quad \text{as } t \to \infty$$

for μ-almost every $x \in X$.

4. The family v_t satisfies the $(L^p(\mu), L^r(\mu))$ *exponential mean ergodic theorem*, $p \geq r$, if there exist $t_0 \geq 0$, $C_{p,r} > 0$, and $\theta_{p,r} > 0$ such that for every $f \in L^p(\mu)$ and $t \geq t_0$,

$$\left\| \pi_X(v_t)f - \int_X f \, d\mu \right\|_{L^r(\mu)} \leq C_{p,r} e^{-t\theta_{p,r}} \|f\|_{L^p(\mu)}.$$

5. The family v_t satisfies the $(L^p(\mu), L^r(\mu))$ *exponential strong maximal inequality*, $p \geq r$, if there exist $t_0 \geq 0$, $C_{p,r} > 0$, and $\theta_{p,r} > 0$ such that for every $f \in L^p(\mu)$,

$$\left\| \sup_{t \geq t_0} e^{t\theta_{p,r}} \left| \pi_X(v_t)f - \int_X f \, d\mu \right| \right\|_{L^r(\mu)} \leq C_{p,r} \|f\|_{L^p(\mu)}.$$

6. The family v_t satisfies the $(L^p(\mu), L^r(\mu))$ *exponential pointwise ergodic theorem*, $p \geq r$, if there exist $t_0 \geq 0$ and $\theta_{p,r} > 0$ such that for every $f \in L^p(\mu)$ and $t \geq t_0$,

$$\left| \pi_X(v_t)f(x) - \int_X f \, d\mu \right| \leq B_{p,r}(x, f)e^{-t\theta_{p,r}} \quad \text{for } \mu\text{-a.e. } x \in X,$$

with the estimator $B_{p,r}(x, f)$ satisfying the norm estimate

$$\|B_{p,r}(\cdot, f)\|_{L^r(\mu)} \leq C_{p,r} \|f\|_{L^p(\mu)}.$$

Remark 3.2. The main motivation to consider the exponential strong maximal inequality in $(L^p(\mu), L^r(\mu))$ is that it implies the exponential pointwise ergodic theorem in $(L^p(\mu), L^r(\mu))$, as well as exponentially fast norm convergence to the ergodic mean.

We recall, in comparison, that the ordinary strong maximal inequality implies pointwise convergence almost everywhere only after we have also established the existence of a dense subspace where almost sure pointwise convergence holds. In addition, convergence in norm requires a separate further argument.

Remark 3.3. If the mean ergodic theorem holds in $L^p(\mu)$, using appoximation by bounded functions and Hölder inequality, one can deduce the mean ergodic theorem in $L^q(\mu)$ for $1 \leq q \leq p$. Similarly, the strong maximal inequality (respectively, the exponential mean ergodic theorem, the exponential strong maximal inequality) in $(L^p(\mu), L^r(\mu))$ implies the strong maximal inequality (respectively, the exponential mean ergodic theorem, the exponential strong maximal inequality) in $(L^q(\mu), L^s(\mu))$ for $q \geq p$ and $1 \leq s \leq r$.

3.2 S-ALGEBRAIC GROUPS AND UPPER LOCAL DIMENSION

We now define the class of S-algebraic groups which will be our main focus.

Definition 3.4. S-algebraic groups.

1. Let F be a locally compact nondiscrete field and let G be the group of F-points of a semisimple linear algebraic group defined over F, with positive F-rank (namely, containing an F-split torus of positive dimension over F). We assume in addition that G is algebraically connected and does not have nontrivial anisotropic (i.e., compact) algebraic factor groups defined over F. We also assume, for simplicity, that G^+ is of finite index in G (see §3.8.2).

2. By an S-algebraic group we mean any finite product of the groups described in the preceding paragraph.

The unitary representation theory of S-algebraic groups has a number of useful features which we will use extensively below.

Another property of S-algebraic groups which is crucial for handling their lattice points is the finiteness of their upper local dimension as defined by natural choices of neighborhood bases. Let us introduce the following.

Definition 3.5. Upper local dimension. A family $\{\mathcal{O}_\varepsilon\}_{0<\varepsilon<1}$ of neighborhoods of e in an lcsc group G such that $\mathcal{O}_\varepsilon$ are symmetric, bounded, and decreasing with ε has *finite upper local dimension* if

$$\varrho_0 \stackrel{\text{def}}{=} \limsup_{\varepsilon \to 0^+} \frac{\log m_G(\mathcal{O}_\varepsilon)}{\log \varepsilon} < \infty. \tag{3.2}$$

Remark 3.6.

1. When M is a Riemannian manifold and $\mathcal{O}_\varepsilon$ are the balls w.r.t. the Riemannian metric, the condition $m_M(\mathcal{O}_\varepsilon) \geq C_\rho \varepsilon^\rho$, $\varepsilon > 0$, is equivalent to $\dim(M) \leq \rho$.

2. When G is an S-algebraic group, we will always take $\mathcal{O}_\varepsilon$ to be the sets $\mathcal{U}_\varepsilon \times K_0$, where $\mathcal{U}_\varepsilon$ is the family of Riemannian balls in the Archimedean component of G (if it exists) and K_0 is a fixed compact open subgroup of the totally disconnected component of G. Thus the local dimension of $\mathcal{O}_\varepsilon$ is the dimension of the Archimedean component.

3.3 ADMISSIBLE AND COARSELY ADMISSIBLE SETS

We begin our discussion of admissibility by introducing a coarse version of it, which will be useful in what follows.

Definition 3.7. Coarse admissibility. Let G be an lcsc group with left Haar measure m_G. An increasing family of bounded Borel subsets G_t ($t \in \mathbb{R}_+$ or $t \in \mathbb{N}_+$) of G will be called *coarsely admissible* if

- For every bounded $B \subset G$, there exists $c = c_B > 0$ such that for all sufficiently large t,

$$B \cdot G_t \cdot B \subset G_{t+c}. \tag{3.3}$$

- For every $c > 0$, there exists $D > 0$ such that for all sufficiently large t,

$$m_G(G_{t+c}) \leq D \cdot m_G(G_t). \tag{3.4}$$

It will be important in our considerations later on that coarse admissibility implies some volume growth for our family G_t, provided that the group is compactly generated. This property will play a role in the spectral estimates that will arise in the proofs of Theorems 4.1 and 4.2. Thus let us note the following.

Proposition 3.8. Coarse admissibility implies volume growth. *When G is compactly generated, coarse admissiblity for an increasing family of bounded Borel subsets $G_t, t > 0$, of G implies that for any bounded symmetric generating set S of G, there exist $a = a(S) > 0$ and $b = b(S) \geq 0$ such that $S^n \subset G_{an+b}$.*

Proof. Let S be a compact symmetric generating set. Taking B to be a bounded open set containing the identity together with $G_{t_0} \cup G_{t_0}^{-1}$ and applying condition (3.3), we conclude that G_{t_0+c} contains an open neighborhood of the identity. Then, assuming without loss of generality that $e \in S$, we have $S \subset SG_{t_0+c}S \subset G_{t_1}$. Applying condition (3.3) repeatedly, we conclude that $S^n \subset G_{t_1+nc_1}$, and the required property follows. $\square$

Remark 3.9. Sequences in totally disconnnected groups. If G is totally disconnected and $K \subset G$ is a compact open subgroup, then G/K is a discrete countable metric space. If $G_t \subset G, t \in \mathbb{R}_+$, is an increasing family of bounded sets, then their projections to G/K will yield only a sequence of distinct sets. Since it is the large-scale behavior of the sets that we are mostly interested in, it is natural to assume that in the totally disconnected case the family G_t is in fact countable, and we then parametrize it by $G_t, t \in \mathbb{N}_+$. This convention will greatly simplify our notation below.

We now consider the following abstract notion of admissible families, which (as we shall see) generalizes the one introduced in §1.1.

Definition 3.10. Admissible families.

1. *Admissible one-parameter families.* Let G be an lcsc group, fix a family of neighborhoods $\{\mathcal{O}_\varepsilon\}_{0<\varepsilon<1}$ of e in G such that $\mathcal{O}_\varepsilon$ are symmetric, bounded, and decreasing with ε.

   An increasing one-parameter family of bounded Borel subsets $G_t, t \in \mathbb{R}_+$, on an lcsc group G will be called *admissible* (w.r.t. to the family $\mathcal{O}_\varepsilon$) if it is coarsely admissible and there exist $c > 0, t_0 > 0$, and $\varepsilon_0 > 0$ such that for $t \geq t_0$ and $0 < \varepsilon \leq \varepsilon_0$,

$$\mathcal{O}_\varepsilon \cdot G_t \cdot \mathcal{O}_\varepsilon \subset G_{t+c\varepsilon}, \tag{3.5}$$

$$m_G(G_{t+\varepsilon}) \leq (1 + c\varepsilon) \cdot m_G(G_t). \tag{3.6}$$

2. *Admissible sequences.* An increasing sequence of bounded Borel subsets G_t, $t \in \mathbb{N}_+$, on an lcsc totally disconnected group G will be called *admissible* if it is coarsely admissible and there exist $t_0 > 0$ and a compact open subgroup K_0 such that for $t \geq t_0$,

$$K_0 G_t K_0 = G_t . \tag{3.7}$$

Let us note the following regarding admissibility.

Remark 3.11.

1. When G is connected and $\mathcal{O}_\varepsilon$ are Riemannian balls, every $\mathcal{O}_\varepsilon$ generates G, and so it is clear that admissibility of the one-parameter family G_t implies coarse admissibility (and thus also volume growth, see Proposition 3.8). However, this argument fails for S-algebraic groups which have a totally disconnected simple component, and so we have required coarse admissibility explicitly in the definition.

2. Condition (3.6) is of course equivalent to the function $\log m_G(G_t)$ being *uniformly* locally Lipschitz-continuous for sufficiently large t. Furthermore, note that

$$\|\beta_{t+\varepsilon} - \beta_t\|_{L^1(G)} = \int_G \frac{|m_G(G_t)\chi_{G_{t+\varepsilon}} - m_G(G_{t+\varepsilon})\chi_{G_t}|}{m_G(G_t) \cdot m_G(G_{t+\varepsilon})} dm_G$$

$$= \frac{2(m_G(G_{t+\varepsilon}) - m_G(G_t))}{m_G(G_{t+\varepsilon})} .$$

It follows that admissibility implies that the map $t \mapsto \beta_t$ is uniformly locally Lipschitz-continuous as a map from $[t_0, \infty)$ to the Banach space $L^1(G)$. The converse also holds, provided we assume in addition that the ratio of $m_G(G_{t+\varepsilon})$ and $m_G(G_t)$ is uniformly bounded for $t \geq t_0$ and $\varepsilon \leq \varepsilon_0$.

It is natural to also introduce the following more general condition, which will play an important role below.

Definition 3.12. Hölder-admissible families. The family G_t will be called *Hölder-admissible* provided it satisfies, for some $0 < a \leq 1, c > 0, t \geq t_0$, and $0 < \varepsilon \leq \varepsilon_0$, the volume Hölder continuity condition

$$m_G(G_{t+\varepsilon}) \leq (1 + c\varepsilon^a) \cdot m_G(G_t) \tag{3.8}$$

and the Hölder stability condition

$$\mathcal{O}_\varepsilon \cdot G_t \cdot \mathcal{O}_\varepsilon \subset G_{t+c\varepsilon^a}. \tag{3.9}$$

3.4 ABSOLUTE CONTINUITY AND EXAMPLES OF ADMISSIBLE AVERAGES

We show that admissible one-parameter families satisfy an absolute continuity property, which will be crucial in the proof of ergodic theorems in the absence of a spectral gap, and thus in Theorems 4.1 and 4.5.

To decribe the property, let us first note that a one-parameter family G_t gives rise to the gauge $|\cdot| : G \to [t_0, \infty)$ defined by $|g| = \inf \{s \geq t_0 ;\ g \in G_s\}$. If the family G_t satisfies the condition $\cap_{r>t} G_r = G_t$ for every $t \geq t_0$, then conversely the family G_t is determined by the gauge, namely, $G_t = \{g \in G ;\ |g| \leq t\}$ for $t \geq t_0$. Note that admissibility implies that $\cap_{r>t} G_r$ can differ from G_t only by a set of measure zero if $t \geq t_0$. Clearly, the resulting family is still admissible, and so we can and will assume from now on that the family G_t is indeed determined by its gauge.

Proposition 3.13. Absolute continuity. *An admissible one-parameter family G_t (w.r.t. a basis $\mathcal{O}_\varepsilon, 0 < \varepsilon < \varepsilon_0$) on an lcsc group G has the following property. The map $g \mapsto |g|$ from G to $[t_0, \infty)$ given by the associated gauge maps the Haar measure on G to a measure on $[t_0, \infty)$, and its restriction (t_0, ∞) is absolutely continuous with respect to the linear Lebesgue measure.*

Proof. The measure η induced on $\mathbb{R}_+$ by the map $g \mapsto |g|$ is by definition $\eta(J) = m_G(\{g \in G ;\ |g| \in J\})$ for any Borel set $J \subset \mathbb{R}_+$. Let us assume that $J \subset (t_0, t_1)$ and that $\ell(J) = 0$ (namely, J has linear Lebesgue measure zero) and show that $\eta(J) = 0$. Indeed, for any $\kappa > 0$ there exists a covering of J by a sequence of intervals I_i, with $\sum_{i=1}^{\infty} \ell(I_i) < \kappa$. Subdividing the intervals if necessary, we can assume that $\ell(I_i) < \varepsilon_0$. By (3.6), for all $\varepsilon < \varepsilon_0$ and $t \in (t_0, t_1)$,

$$\eta((t, t + \varepsilon]) = m_G(\{g ;\ t < |g| \leq t + \varepsilon\}) = m_G(G_{t+\varepsilon}) - m_G(G_t)$$
$$\leq c \varepsilon m_G(G_t) \leq c m_G(G_{t_1}) \ell((t, t + \varepsilon]).$$

Denoting $c m_G(G_{t_1})$ by C, we see that

$$\eta(J) \leq \sum_{i=1}^{\infty} \eta(I_i) \leq C \sum_{i=1}^{\infty} \ell(I_i) \leq C\kappa,$$

and since κ is arbitrary, it follows that $\eta(J) = 0$ and thus η is absolutely continuous w.r.t. ℓ. $\square$

Let us now verify the first assertion made regarding admissibility in §1.1. The second assertion is discussed immediately below, and the third is proved in Lemma 5.24.

Proposition 3.14. *When G is a connected Lie group and $\mathcal{O}_\varepsilon$ are the balls defined by a left-invariant Riemannian metric, the property of admissibility is independent of the Riemannian metric chosen to define it (but the constant c may change).*

Proof. We need only verify that (3.5) is still satisfied, possibly with another constant c (but keeping ε_0 and t_0 the same) if we choose another Riemannian metric. Fix such a Riemannian metric and denote its balls by $\mathcal{O}'_\varepsilon$. First, note that it suffices to verify (3.5) for all $\varepsilon < a$, where a is *any* positive constant. Indeed, then for $t \geq t_0$ and $\varepsilon < a$,

$$\mathcal{O}'_{2\varepsilon} G_t \mathcal{O}'_{2\varepsilon} = \mathcal{O}'_\varepsilon \mathcal{O}'_\varepsilon G_t \mathcal{O}'_\varepsilon \mathcal{O}'_\varepsilon \subset \mathcal{O}'_\varepsilon G_{t+c'\varepsilon} \mathcal{O}'_\varepsilon \subset G_{t+2c'\varepsilon}.$$

Here we have used the property $\left(\mathcal{O}'\right)^n_\varepsilon = \mathcal{O}'_{n\varepsilon}$, which is valid for balls with respect to an invariant Riemannian metric. It follows that $\mathcal{O}'_\varepsilon G_t \mathcal{O}'_\varepsilon \subset G_{t+c'\varepsilon}$ holds for all $0 < \varepsilon \leq \varepsilon_0$.

Now note that it is possible to choose $a > 0$ small enough so that for $\varepsilon < a$ there exists a fixed m independent of ε such that

$$\mathcal{O}'_\varepsilon \subset \mathcal{O}^m_\varepsilon = \mathcal{O}_{m\varepsilon} .$$

This fact follows by applying the exponential map in a sufficiently small ball in the Lie algebra of G and using the fact that any two norms on the Lie algebra are equivalent. It then follows that for $\varepsilon < a, t \geq t_0$,

$$\mathcal{O}'_\varepsilon G_t \mathcal{O}'_\varepsilon \subset \mathcal{O}^m_\varepsilon G_t \mathcal{O}^m_\varepsilon \subset G_{t+mc\varepsilon}$$

as required, with $c' = mc$. $\square$

Admissible averages exist in abundance on S-algebraic groups. We refer the reader to §7.1 for a proof of the following result.

Theorem 3.15. *For an S-algebraic group $G = G(1)\cdots G(N)$ as in Definition 3.4, the following families of sets $G_t \subset G$ are admissible, where a_i are any positive constants, and $t \in \mathbb{R}_+$ when S contains at least one infinite place and $t \in \mathbb{N}_+$ otherwise.*

1. *Let S consist of infinite places and let $G(i)$ be a closed subgroup of the isometry group of a symmetric space $X(i)$ of nonpositive curvature equipped with the Cartan-Killing metric. Let d_i be the associated distance, and for $u_i, v_i \in X(i)$, define*

$$G_t = \{(g_1, \ldots, g_N) : \sum_i a_i d_i (u_i, g_i \cdot v_i) < t\}.$$

2. *Let S consist of infinite places and let $\rho_i : G(i) \to \mathrm{GL}(V_i)$ be proper rational representations. For norms $\|\cdot\|_i$ on $\mathrm{End}(V_i)$, define*

$$G_t = \{(g_1, \ldots, g_N) : \sum_i a_i \log \|\rho_i(g_i)\|_i < t\}.$$

3. *For infinite places, let $X(i)$ be the symmetric space of $G(i)$ equipped with the Cartan-Killing distance d_i, and for finite places, let $X(i)$ be the Bruhat-Tits building of $G(i)$ equipped with the path metric d_i on its 1-skeleton. For $u_i \in X(i)$, define*

$$G_t = \{(g_1, \ldots, g_N) : \sum_i a_i d_i (u_i, g_i \cdot u_i) < t\}.$$

4. *Let $\rho_i : G(i) \to \mathrm{GL}(V_i)$ be proper representations, rational over the fields of definition F_i. For infinite places, let $\|\cdot\|_i$ be a Euclidean norm on $\mathrm{End}(V_i)$ and assume that $\rho_i(G(i))$ is self-adjoint: $\rho_i(G(i))^t = \rho_i(G(i))$. For finite places, let $\|\cdot\|_i$ be the $\max$ norm on $\mathrm{End}(V_i)$. Define*

$$G_t = \{(g_1, \ldots, g_N) : \sum_i a_i \log \|\rho_i(g_i)\|_i < t\}.$$

An important class of families are those defined by height functions on S-algebraic groups. In Theorem 7.19, we will establish the following.

Theorem 3.16. Hölder-admissible families. *For an S-algebraic group $G = G(1)$* *$\cdots G(N)$ as in Definition 3.4, let $\rho_i : G(i) \to \mathrm{GL}(V_i)$ be proper representations, rational over the fields of definition F_i. For infinite places, let $\|\cdot\|_i$ be any Euclidean norm on $\mathrm{End}(V_i)$. For finite places, let $\| \cdot \|_i$ be the* max *norm on $\mathrm{End}(V_i)$. Define (for any positive constants a_i)*

$$G_t = \{(g_1, \dots, g_N) : \sum_i a_i \log \|\rho_i(g_i)\|_i < t\}.$$

Then G_t are Hölder-admissible.

3.5 BALANCED AND WELL-BALANCED FAMILIES
ON PRODUCT GROUPS

In any discussion of ergodic theorems for averages ν_t on a product group $G = G(1) \times G(2)$, it is necessary to discuss the behavior of the two projections ν_t^1 and ν_t^2 to the factor groups. Indeed, consider the case where one of these projections, say ν_t^1, assigns a fixed fraction of its measure to a bounded set for all t. Then choosing an ergodic action of $G(1)$, we can view it as an ergodic action of the product in which $G(2)$ acts trivially, and it is clear that the ergodic theorems will fail for ν_t in this action. Thus it is necessary to require one of the following two conditions. Either the projections of the averages ν_t to the noncompact factors do not assign a fixed fraction of their measures to a bounded set, or alternatively that the action is irreducible, namely, every noncompact factor acts ergodically. This unavoidable assumption is reflected in the following definitions.

Definition 3.17. Balanced and well-balanced averages. Let $G = G(1) \cdots G(N)$ be an almost direct product of N noncompact compactly generated subgroups. For a set I of indices $I \subset [1, N]$, let J denote its complement and $G(I) = \prod_{i \in I} G(i)$. Let G_t be an increasing family of sets contained in G.

1. G_t will be called *balanced* if for every I satisfying $0 < |I| < N$ and every compact set Q contained in $G(I)$,

$$\lim_{t \to \infty} \frac{m_G(G_t \cap G(J) \cdot Q)}{m_G(G_t)} = 0.$$

2. An admissible family G_t will be called *well balanced* if there exist $a > 0$ and $\eta > 0$ such that for all I satisfying $0 < |I| < N$,

$$\frac{m_G(G_{an} \cap G(J) \cdot S_I^n)}{m_G(G_{an})} \leq Ce^{-\eta n},$$

where S_I is a compact generating set of $G(I)$ consisting of products of compact generating sets of its component groups.

The condition of being well balanced is independent of the choices of compact generating sets in the component groups, but the various constants may change. An explicit sufficient condition for a family of sets G_t defined by a norm on a semisimple Lie group to be well balanced is given in §8.3.

Let us note the following important natural examples of admissible well-balanced families of averages and state an estimate of their boundary measures which will play an important role in the proof of Theorem 4.1. A complete proof of Theorem 3.18 will be given in Chapter 7.

Observe that when G_t is an admissible family, the Haar measure m_G can be disintegrated as $m_G = m_G|_{G_{t_0}} + \int_{t_0}^{\infty} m_t \, dt$, where m_t is a measure supported on ∂G_t for almost every t. Indeed, the disintegration formula holds by Proposition 3.13. We can now state the following.

Theorem 3.18. *Let $G = G(1) \cdots G(N)$ be an S-algebraic group and let ℓ_i denote the standard $CAT(0)$-metric on either the symmetric space $X(i)$ or the Bruhat-Tits building $X(i)$ associated to $G(i)$. For $1 < p < \infty$ and $u_i \in X(i)$, define*

$$G_t = \{(g_1, \ldots, g_N) : \sum_i \ell_i(u_i, g_i u_i)^p < t^p\}.$$

1. *There exist $\alpha, \beta > 0$ such that for every nontrivial projection $G \to L$,*

$$m_G(G_t \cap \mathrm{proj}^{-1}(L_{\alpha t})) \ll e^{-\beta t} \cdot m_G(G_t),$$

   namely, the averages are well balanced ($t \geq t_0$).

2. *If G has at least one Archimedean factor and $p \leq 2$, then the family G_t is admissible and there exist $\alpha, \beta > 0$ such that for every nontrivial projection $G \to L$ for almost every t,*

$$m_t(\partial G_t \cap \mathrm{proj}^{-1}(L_{\alpha t})) \ll e^{-\beta t} \cdot m_t(\partial G_t),$$

   such averages will be called boundary-regular ($t > t_0$).

Let us introduce the following definition, which was referred to in Theorems 1.2 and 1.10.

Definition 3.19. Standard averages. Let G be an S-algebraic group as in Definition 3.4 and represent G as a product $G = G(1) \cdots G(N)$ of its simple components. We will refer to any of the families defined in Theorems 3.15, 3.16, and 3.18 as *standard averages*.

If the family in addition satisfies the estimate in Theorem 3.18(2), it will be called *boundary-regular*.

3.6 ROUGHLY RADIAL AND QUASI-UNIFORM SETS

We now define several other stability properties for families of sets G_t that will be useful in the arguments below.

Definition 3.20. Quasi-uniform families. An increasing one-parameter family of bounded Borel subsets G_t, $t > 0$, of G will be called *quasi-uniform* if it satisfies the following two conditions:

- Quasi-uniform local stability. For every $\varepsilon > 0$, there exists a neighborhood $\mathcal{O}$ of e in G such that for all sufficiently large t,

$$\mathcal{O} \cdot G_t \subset G_{t+\varepsilon}. \tag{3.10}$$

- Quasi-uniform continuity. For every $\delta > 0$, there exists $\varepsilon > 0$ such that for all sufficiently large t,

$$m_G(G_{t+\varepsilon}) \leq (1+\delta) \cdot m_G(G_t). \tag{3.11}$$

Note that (3.11) is equivalent to the function $\log m_G(G_t)$ being quasi-uniformly continuous in t and implies that $t \mapsto \beta_t$ is quasi-uniformly continuous in the $L^1(G)$-norm. If the ratio of $m_G(G_{t+\varepsilon})$ and $m_G(G_t)$ is uniformly bounded for $0 < \varepsilon \leq \varepsilon_0$, the converse holds as well.

An important ingredient in our analysis below will be the existence of a radial structure on the groups under consideration. Thus let G be an lcsc group and K a compact subgroup. Sets which are bi-invariant under left and right translations by K will be used in order to dominate sets which are not necessarily bi-K-invariant.

In particular, we shall utilize special bi-K-invariant sets, called ample sets, which play a key role in the ergodic theorems proved in [N5], which we will use below. We recall the definitions.

Definition 3.21. Roughly radial sets and ample sets. Let $K \subset G$ be a fixed compact subgroup, which is of finite index in a maximal compact subgroup of G. Let $\mathcal{O}$ be a fixed neighborhood of $e \in G$ and C, D positive constants.

1. $B \subset G$ is called *left-radial* (or more precisely left K-radial) if it satisfies $KB = B$.

2. (See [N4].) A measurable set $B \subset G$ of positive finite measure is called *roughly radial* (or more precisely (K, C)-radial), provided that

$$m_G(KBK) \leq Cm_G(B).$$

3. (See [N5].) A measurable set $B \subset G$ of positive finite measure is called *ample* (or more precisely $(\mathcal{O}, D, K)$-ample) if it satisfies

$$m_G(K\mathcal{O}BK) \leq Dm_G(B).$$

To illustrate the definition of ampleness, first consider the case where G is a connected semisimple Lie group. We can fix a maximal compact subgroup K of G and consider the symmetric space $S = G/K$, with the distance d derived from the Riemannian metric associated with the Killing form. Ampleness can be equivalently defined as follows. For a K-invariant set $B \subset G/K$, consider the r-neighborhood of B in the symmetric space given by

$$U_r(B) = \{gK \in G/K \,;\, d(gK, B) < r\}.$$

Then B is $(\mathcal{O}_r, D, K)$-ample iff $m_{G/K}(U_r(B)) \leq Dm_{G/K}(B)$, where $\mathcal{O}_r$ is the lift to G of a ball of radius r and center K in G/K.

The following simple facts are obvious from the definition, but since they will be used below, we record them for completeness.

Proposition 3.22. *Any family of coarsely admissible sets on an lcsc group is (K, C)-radial for some finite C and any maximal compact subgroup K, as well as $(\mathcal{O}, D, K)$-ample, for some neighborhood $\mathcal{O}$ and $D > 0$.*

Proof. By the definition of coarse admissibility, for the compact set K there exists $c \geq 0$ with $K \cdot G_t \cdot K \subset G_{t+c}$. Therefore $m_G(KG_tK) \leq Dm_G(G_t)$, so that the family G_t is (K, D)-radial. The fact that G_t are ample sets is proved in the same way. $\qquad\square$

Let us note that when G is totally disconnected and there exists a compact open subgroup $Q \subset \mathcal{O}_\varepsilon$ satisfying $QG_tQ = G_t$ for all $t \geq t_0$, then G_t are (K, C)-radial for any choice of a maximal compact subgroup (e.g., a good maximal compact subgroup; see [T] and [Si]). Indeed, Q is of finite index in K, and denoting the index by N, we have

$$KG_tK = \cup_{i,j=1}^{N}k_iQG_tQk'_j \subset \cup_{i,j=1}^{N}k_iG_tk'_j.$$

It follows that $m_G(KG_tK) \leq N^2m_G(G_t)$ and G_t is (K, N^2)-radial.

3.7 SPECTRAL GAP AND STRONG SPECTRAL GAP

We begin by recalling the standard definition of a spectral gap, as follows.

Definition 3.23. Spectral gap.

1. A strongly continuous unitary representation π of an lcsc group G is said to have a spectral gap if $\|\pi(\mu)\| < 1$ for some (or equivalently, all) absolutely continuous symmetric probability measure μ whose support generates G as a group.

2. Equivalently, π has a spectral gap if the underlying Hilbert space does not admit an asymptotically G-invariant sequence of unit vectors, namely, a sequence satisfying $\lim_{n\to\infty} \|\pi(g)v_n - v_n\| = 0$ uniformly on compact sets in G.

3. A measure-preserving action of G on a σ-finite measure space (X, m) is said to have a spectral gap if the unitary representation π_X^0 of G in the space orthogonal to the space of G-invariant functions has a spectral gap. Thus in the case of an ergodic probability measure–preserving action, the representation in question is on the space $L_0^2(X)$ of function of zero integral.

4. An lcsc group G is said to have *property T* [Ka] provided that every strongly continuous unitary representation which does not have G-invariant unit vectors has a spectral gap.

If $G = G(1)G(2)$ is an (almost) direct product group, and there does not exist a sequence of unit vectors which is asymptotically invariant under every $g \in G$, it may still be the case that there exists such a sequence asymptotically invariant under the elements of a subgroup of G, for example, $G(1)$ or $G(2)$. It is thus natural to introduce the following definition, which plays an important role in our discussion.

Definition 3.24. Strong spectral gaps. Let $G = G(1) \cdots G(N)$ be an almost direct product of N lcsc subgroups. A strongly continuous unitary representation π of G has a strong spectral gap (w.r.t. the given decomposition) if the restriction of π to every almost direct factor $G(i)$ has a spectral gap.

Remark 3.25.

1. Let $G(1) = G(2) = SL_2(\mathbb{R})$, $G = G(1) \times G(2)$, and $\pi = \pi_1 \otimes \pi_2$, where π_1 has a spectral gap and π_2 does not (but has no invariant unit vectors). It is possible to construct two admissible families G_t and G'_t on G such that $\|\pi(\beta_t)\| \leq Ce^{-\theta t}$ but $\|\pi(\beta'_t)\| \geq b > 0$. For example, G_t can be taken as the inverse images of the families of balls of radius t in $\mathbb{H} \times \mathbb{H}$ w.r.t. the Cartan-Killing metric. For a construction of G'_t, one can use (in the obvious way) the nonbalanced averages constructed in §8.2.

2. In order to obtain conclusions which assert that the mean, maximal, or pointwise theorem holds for β_t with an exponential rate, it is of course necessary that in the representation π_X^0 in $L_0^2(X, \mu)$ β_t have the exponential decay property, namely, $\|\pi_X^0(\beta_t)\| \leq Ce^{-\theta t}, \theta > 0$. In Theorem 5.11 we will give sufficient conditions for the latter property to hold.

3. Consider the special case $X = G/\Gamma$, where G is a semisimple Lie group and Γ is a lattice subgroup. It is a standard corollary of the theory of elliptic operators on compact manifolds that if Γ is cocompact, then the (positive) Laplacian Δ on G/Γ has a spectral gap above zero, namely, $\|\exp(-\Delta)\| < 1$. It then follows that the G-action on $L_0^2(G/\Gamma)$ has a spectral gap. It was shown that the same holds for any lattice, including nonuniform ones (see [BoGa] and [Be, Lem. 3]).

4. If Γ is an irreducible lattice and G has no nontrivial compact factors, then $L_0^2(G/\Gamma)$ has a strong spectral gap (see [KS] for a recent discussion). However, the strong spectral gap property may fail in general ergodic actions. This motivated the formulation of our results in a way which makes the dependence on the strong spectral gap explicit; namely, this assumption is redundant for averages which are well balanced.

3.8 FINITE-DIMENSIONAL SUBREPRESENTATIONS

3.8.1 Mixing

In order to state the ergodic theorems, we will need the notion of totally weak-mixing actions:

Definition 3.26. Totally weak-mixing. A measure-preserving action of a group G on a probability space (X, μ) is called

1. *weak-mixing* if the unitary representation in $L_0^2(X)$ does not contain nontrivial finite-dimensional subrepresentations.

2. *totally weak-mixing* if, when $G = G(1)G(2) \cdots G(N)$ is an almost direct product of N normal subgroups, the only finite-dimensional subrepresentation that $G(i)$ admits is the trivial one (possibly with multiplicity greater than 1), or equivalently, if in the space orthogonal to the $G(i)$-invariants no finite-dimensional representations of $G(i)$ occur, $1 \leq i \leq N$.

We will also apply below the obvious generalizations of these notions to general unitary representations.

To demonstrate the necessity of this condition, it suffices to consider $G = PGL_2(\mathbb{Q}_p)$ and note that it admits a continuous character χ_2 onto $\mathbb{Z}_2 = \{\pm 1\}$. It is easily seen that for the natural radial averages β_n on G (projecting onto the balls on the Bruhat-Tits tree), the sequence $\chi_2(\beta_n)$ *does not* converge at all. $\chi_2(\beta_{2n})$ does in fact converge but not to the ergodic mean. Thus, in general, the limiting value, if it exists, of $\tau(\beta_t)$ (or subsequences thereof) in finite-dimensional representations τ must be incorporated explicitly into the formulation of the ergodic theorems for G. We refer the reader to [N6, §10.5] for a fuller discussion.

Furthermore, note that obviously the ergodic action of G on the two-point space $G/\ker \chi_2$ has a spectral gap, so weak-mixing is essential even when a spectral gap is present.

3.8.2 The group G^+

It will be crucial for the validity of ergodic theorems to exclude finite-dimensional representations in the spectrum. For this purpose, we recall the properties of the normal subgroup denoted G^+ of G as follows.

1. Consider an algebraic group G defined over a local field F which is F-isotropic, almost simple, and algebraically connected as in Definition 3.4. Then G contains a canonical cocompact normal subgroup denoted G^+, which can be defined as the group generated by the unipotent radicals of a pair of opposite minimal parabolic F-subgroups of G (see, e.g., [M, §1.5 and §2.3] for a discussion).

   We recall that if $F = \mathbb{C}$, then $G^+ = G$, and when $F = \mathbb{R}$, G^+ is the connected component of the identity in the Hausdorff topology (which is of finite index in G). When the characteristic of F is zero, $[G : G^+] < \infty$. We note that it is often the case that $G = G^+$ even in the totally disconnected case. Thus when G is simply connected and almost F-simple, then $G^+ = G$, and this includes, for example, the groups $SL_n(F)$ and $Sp_{2n}(F)$ (see, e.g., [M, §1.4 and §2.3]).

2. A key property of G^+ is that it does not admit any proper finite index subgroup (see, e.g., [M, Cor. 1.5.7]). As a result, it follows that G^+ does not admit any nontrivial finite-dimensional unitary representations. Put otherwise, an irreducible unitary representation of G is finite-dimensional if and only if it admits a G^+-invariant unit vector. In particular, every ergodic action of G^+ is weak-mixing, and if it is irreducible, each component is weak-mixing.

3. We note the following fact: when $[G : G^+] < \infty$, clearly every irreducible nontrivial unitary representation of G^+ appears as a subrepresentation of the representation of G obtained from it by induction. The induced representation has no G^+-invariant unit vectors, and hence its matrix coefficients satisfy the estimates that irreducible infinite-dimensional unitary representations of G without G^+-invariant unit vectors satisfy. In particular, the K-finite matrix coefficients are in $L^p(G^+)$ for some $p < \infty$ (see Theorem 5.6).

Chapter Four

Main results and an overview of the proofs

In the present chapter we will state our main results, namely, the ergodic theorems for actions of G and for actions of Γ in the presence of a spectral gap and in the absence of a spectral gap. We will also give an overview of the proofs of these results, as well as an overview of the proofs of the volume regularity results that are the subject of Chapter 7.

4.1 STATEMENT OF ERGODIC THEOREMS FOR S-ALGEBRAIC GROUPS

Let us now formulate the two basic ergodic theorems for actions of an S-algebraic group G, which we will prove in the following chapter. As usual, it is the maximal and exponential-maximal inequalities that will serve as our main technical tool in the proof of the ergodic theorems. The maximal inequalities will also be essential later on in establishing the connection between the averages on the group and those on the lattice.

Below we will use the notation and terminology established in Chapters 1 and 3. In the absence of a spectral gap, we have the following.

Theorem 4.1. Ergodic theorems in the absence of a spectral gap. *Let G be an S-algebraic group as in Definition 3.4. Let (X, μ) be a totally weak-mixing action of G and let $\{G_t\}$ be a coarsely admissible one-parameter family or sequence. Then the Haar-uniform averages β_t on G_t satisfy the strong maximal inequality in (L^p, L^r) for $p \geq r \geq 1$, $(p, r) \neq (1, 1)$. Furthermore, if the G-action is irreducible, β_t satisfy*

1. the mean ergodic theorem in L^p for $1 \leq p < \infty$.

2. the pointwise ergodic theorem in L^p for $1 < p < \infty$, provided G_t are admissible and left-radial.

The conclusions still hold when the action is reducible, provided that G_t are admissible, and the averages β_t are left-radial and balanced (for the mean theorem) or standard, well balanced, and boundary-regular (for the pointwise theorem).

We note that by Theorem 3.18, many natural radial averages do indeed satisfy all the conditions required in Theorem 4.1 and thus satisfy the pointwise ergodic theorem even in reducible actions.

In the presence of a spectral gap, we have the following.

Theorem 4.2. Ergodic theorems in the presence of a spectral gap. *Let G be an S-algebraic group as in Definition 3.4. Let (X, μ) be a totally weak-mixing action of G on a probability space. Let G_t be a Hölder-admissible one-parameter family or an admissible sequence when S consists of finite places. Assume either that the representation of G on $L_0^2(X)$ has a strong spectral gap or that it has a spectral gap and the family G_t is well balanced. Then the averages β_t satisfy*

1. *the exponential mean ergodic theorem in (L^p, L^r) for $p \geq r \geq 1$, provided $(p, r) \neq (1, 1)$ and $(p, r) \neq (\infty, \infty)$.*

2. *the exponential strong maximal inequality in (L^p, L^r) for $p > r \geq 1$.*

3. *the exponential pointwise ergodic theorem in (L^p, L^r) for $p > r \geq 1$.*

We note that by Theorems 3.15 and 3.18, Hölder-admissible well-balanced families do exist in abundance.

Let us make the following comments regarding the necessity of the assumptions in Theorems 4.1 and 4.2.

Remark 4.3. On the exponential decay of operator norms.

1. When G is a product of simple groups but is not simple, the assumption that G_t is balanced in Theorem 4.1 and well balanced in Theorem 4.2 is obviously necessary in both cases. To demonstrate this, in the first case we can simply take an ergodic action of $G = G(1) \times G(2)$ which is trivial on one factor. In the second, we can take an ergodic action with a spectral gap for G but such that one of the factors admits an asymptotically invariant sequence of unit vectors of zero integral.

2. In general, it will be seen below that the only property needed to prove Theorem 4.2 for a one-parameter Hölder-admissible family acting in $L_0^2(X)$ is the exponential decay of the operator norms: $\left\| \pi_X^0(\beta_t) \right\| \leq C e^{-\theta t}$. It will be proved in Theorem 5.11 below that this estimate holds for totally weak-mixing actions under the strong spectral gap assumption or when the action has a spectral gap and the averages are well balanced.

3. We note that for averages given in explicit geometric form, exponential decay of the operator norms $\left\| \pi_X^0(\beta_t) \right\|$ can often be established directly (see, e.g., [N4, Thm. 6]). The exponential mean ergodic theorem for averages on G holds in much greater generality and does not require admissibility. This fact is very useful in the solution of lattice point–counting problems. We refer the reader to [GN] for a full discussion and further applications.

4. Similarly, the radiality assumptions in Theorem 4.1 are made specifically in order to estimate certain spectral expressions that arise in the proof of the pointwise ergodic theorem. At issue is the estimate of $\|\pi(\partial\beta_t)\|$, where $\partial\beta_t$ is a singular probability measure supported on the boundary of G_t. We establish the required estimate for left-radial admissible averages in irreducible actions or for standard well-balanced averages in reducible actions. This accounts for the statement of Theorem 4.1.

Remark 4.4. Finite-dimensional representations.

1. The assumption of total weak-mixing is necessary in Theorems 4.1 and 4.2 even for the simple algebraic group $\text{PGL}_2(\mathbb{Q}_p)$, as noted in §3.8.

2. Alternatively, another formulation of Theorem 4.1 for simple algebraic groups (in L^2, say) is that convergence to the ergodic mean (namely, zero) holds for an arbitrary ergodic action when we consider the functions in the orthogonal complement of the space spanned by all finite-dimensional sub-representations. The complete picture requires evaluating the limits of $\tau(\beta_t)$ for finite-dimensional nontrivial representations τ (if they exist!).

3. Similarly, in Theorem 4.2, if the action is such that in fact

$$\|\tau(\beta_t)\| \leq C \exp(-\theta t)$$

for all finite-dimensional nontrivial τ that occur, we still obtain the same conclusion even when finite-dimensional representations are present.

In the following two sections we give an overview of the proof of Theorems 4.1 and 4.2.

4.2 ERGODIC THEOREMS IN THE ABSENCE
OF A SPECTRAL GAP: OVERVIEW

Traditional pointwise ergodic theorems aim to establish that a family of operators $\pi(\beta_t) : L^p(X) \to L^p(X)$ satisfies $\pi(\beta_t)f(x) \to \int_X f d\mu$ for almost all $x \in X$ and $f \in L^p(X)$. As is well known, the usual recipe for the proof of such a result (for L^2-functions, say) involves the following three ingredients:

1. *Mean ergodic theorem:* $\pi_X(\beta_t)f \to \int_X f d\mu$ in the L^2-norm.

2. *Strong maximal inequality:* $\left\|\sup_{t>t_0} |\pi_X(\beta_t)f|\right\|_2 \leq C \|f\|_2$ for some t_0, $C > 0$.

3. *Pointwise convergence on a dense subspace:* $\pi_X(\beta_t)f(x) \to \int_X f d\mu$ almost everywhere for f in a dense subset of $L^2(X)$.

A more general result, namely, the pointwise ergodic theorem in every L^p-space with $p > 1$, may require additional interpolation arguments. Theorem 4.1 (and its corollary Theorem 1.2(1)) state such a traditional ergodic theorem, and we will indeed follow the recipe indicated above in proving it. However, while the results we state are analogous to Wiener's ergodic theorem for actions of $\mathbb{R}^n$ (or more generally to the ergodic theorem for groups of polynomial growth; see [N6]), the methods are necesssarily completely different. In the absence of a Følner sequence, covering arguments, and the transfer principle for amenable groups, we will have to rely on the methods developed for semisimple groups in [N2], [NS2], and [N5]. Spectral methods and geometric comparison arguments on the symmetric spaces play a crucial role here, which we now explain.

4.2.1 The maximal inequality

A simple property of (coarsely) admissible averages β_t is that they are dominated by the corresponding radial averages, namely, the uniform averages on KG_tK, where K is a maximal compact subgroup. Therefore, for the purposes of the maximal inequality, it is sufficient to consider radial averages. It turns out that radial averages satisfy a strong maximal inequality in every L^p, $p > 1$, which is much more general and robust than its Euclidean analog. This fact was established in [N5], and to explain its origin, recall that a connected semisimple Lie group G admits polar coordinates in the form of a Cartan decomposition $G = KAK$, where A is a vector group. Fix positive constants c and C and consider all radial (namely, bi-K-invariant) sets $E \subset G$ whose radial coordinates E_A in A have the following property, which we denote by $A_{(c,C)}$.

Every point $H \in E_A$ is at a distance at most C from a point $H' \in E_A$ such that the ball of radius c with center H' is contained in E_A.

It is clear that any union of subsets of A with this property still satisfies it and that every ball in A with radius c satisfies the property. Thus any subset of A which is a union of c-balls has a radialization that satisfies property $A_{(c,C)}$. Now the strong maximal inequality actually holds for the operator $\sup_E |\pi(\nu_E)f(x)|$, where ν_E is the Haar-uniform measure on E and E ranges over all subsets of G satisfying condition $A_{(c,C)}$, with (c, C) fixed. The underlying cause of this phenomenon has its roots precisely in the exponential volume growth of Riemannian balls on the symmetric space. Namely, exponential volume growth implies that the shell with radii between t and $t + c$ satisfies the same strong maximal inequality satisfied by the ball of radius t (in contrast to the Euclidean situation!). Thus in a product group $G = G(1) \times G(2)$, the strong maximal inequality is satisfied by the product of two arbitrary shells (but of fixed size c) and then also by any set which can be roughly evenly covered by such products. A more elaborate argument taking the volume density and the walls of the Weyl chamber into account is required for simple groups of real rank at least two, but this fact is not strictly necessary here since these groups satisfy property T, so an even stronger maximal inequality holds, as we shall see below.

In fact, the maximal inequality still holds if the supremum is taken over all sets satisfying the more general condition of being ample (see Definition 3.21). In particular, as (coarsely) admissible families consist of ample sets, the ordinary strong maximal inequality follows.

4.2.2 Mean ergodic theorem

The mean ergodic theorem for (coarsely) admissible averages β_t on any lcsc group G certainly follows when the matrix coefficients $\langle \pi_X^0(g)f, f' \rangle$, with $f, f' \in L_0^2(X)$, vanish at infinity. This is the case in all actions of *simple* algebraic groups that we consider since we assume that they are mixing. However, we also consider actions of semisimple and S-algebraic groups, and we specifically admit the possibility that the action may be reducible, namely, that some simple component will not act ergodically. In that case it is of course obvious that the matrix coefficients do not vanish at infinity. The property that $\langle \pi_X^0 (\beta_t) f, f' \rangle \to 0$ then becomes much more delicate and depends on the distribution of mass of the measure β_t and specifically on whether it is balanced among the simple factors. We note, however, that admitting reducible actions of S-algebraic groups is crucial for our later arguments since we intend to prove the ergodic theorems for a lattice action by inducing it to a G-action. There is no general criterion we are aware of that will guarantee that the induced action is irreducible, so that restricting the pointwise ergodic theorem for G to irreducible actions will severely limit the generality of the pointwise ergodic theorem for the lattices in G. We will prove the mean ergodic theorem for balanced admissible families by producing a dense subset of functions $f \in L^2(X)$ where $\pi_X(\beta_t) f(x)$ converges almost everywhere and in L^2-norm to the right limit. It then follows from straightforward approximation arguments that the mean theorem holds in L^2 since it holds on a dense subspace, and then that the mean theorem holds in L^p, $1 \le p < \infty$, if it holds in L^2. This result holds for any (mixing) action of G, but if the action is irreducible, the assumption of balancedness is no longer necessary.

4.2.3 Pointwise convergence on a dense subset

Pointwise convergence of admissible averages on a dense subset in L^2 in the absence of a spectral gap is the most technically demanding and least robust argument among those used in developing the ergodic theorems for averages on G. It requires considerably more stringent assumptions than the maximal inequality theorem and the mean ergodic theorem. This is due to the fact that so far the only viable method that exists for proving pointwise convergence in the presence of exponential volume growth and in the absence of a spectral gap relies on a Sobolev space argument, originating in [N2]. Ultimately, for L^2-functions f which are differentiable vectors in a suitable sense, this method calls for decay estimates of the size of $\left\| \frac{d}{dt} \pi(\beta_t) f \right\|$ as a function of t. The existence of such a derivative together with norm estimates of the associated distribution are quite nontrivial properties, and it is in order to establish these that we must introduce further assumptions. First, for admissible averages we establish that the Lipschitz continuity conditions inherent in their definition suffice to ensure the almost sure existence of $\frac{d}{dt} \beta_t$, which is a signed measure given by a multiple of $\partial \beta_t - \beta_t$. Here $\partial \beta_t$ is the probability measure supported on the boundary ∂G_t of G_t, and the family $\partial \beta_t$ gives rise to the disintegration of the "ball" averages β_t w.r.t. the "sphere" averages $\partial \beta_t$ on ∂G_t. We then use the noncommutative Fourier transform, namely, the unitary representation theory of G, in order to reduce pointwise convergence on a dense subspace to a bound of an appropriate square function associated with the derivative $\frac{d}{dt} \beta_t$. To establish the bound we use the spectral transfer principle to reduce the estimate to the regular representation. Finally, since we assume that the averages β_t and hence $\partial \beta_t$ are left-radial, a favorable estimate can be deduced from the decay properties of the Harish-Chandra Ξ-function, provided the action is irreducible. When the action is reducible, however, the question of whether $\partial \beta_t$ is balanced among the simple factors of G inevitably comes up. For general families β_t this problem seems to be out of reach by current techniques, so in the reducible case we assume further

that the averages are standard and boundary-regular. However, by our analysis of
the regularity of the volume function, many of the most natural averages on any
S-algebraic group do satisfy these assumptions, so the pointwise ergodic theorem
is established for such averages for all (mixing) G-actions.

4.3 ERGODIC THEOREMS IN THE PRESENCE
OF A SPECTRAL GAP: OVERVIEW

Theorem 4.2 (and its corollary Theorem 1.2(2)) differs from traditional ergodic
theorems in several respects. In particular, the mean ergodic theorem is stated in
a quantitative form, and the maximal inequality is replaced by the much stronger
exponential-maximal inequality. In addition, almost everywhere convergence of
$\pi_X(\beta_t)f(x)$ to the ergodic mean occurs at a prescribed exponential rate.

Let us first note that exponentially fast almost sure convergence follows immediately from the exponential strong maximal inequality, which also determines the
rate of convergence. So we are left with the exponential mean ergodic theorem
and the exponential strong maximal inequality, and we now comment on the main
issues that come up in the proof.

4.3.1 Actions with a strong spectral gap

First we assume that the representation π_X^0 is an L^p-representation; namely, $L^2(X)$
contains a dense subspace giving rise to matrix coefficients which are in $L^p(G)$.
This property implies the existence of an integer n such that $\left(\pi_X^0\right)^{\otimes n}$ is isomorphic
to a subrepresentation of a multiple of the regular representation of G. In the setup
of semisimple S-algebraic groups, the two properties are in fact equivalent.

4.3.1.1 The exponential mean ergodic theorem

First, to establish the exponential decay of the norms of $\pi_X^0(\beta_t)$ as operators from
$L_0^2(X)$, we use the spectral transfer principle in [N4], which asserts that $\left\|\pi_X^0(\beta_t)\right\| \leq$
$\left\|\lambda_G(\beta_t)\right\|^{1/n_e}$ for an appropriate integer n_e. Second, establishing that the norms of
the convolution operators $\lambda_G(\beta_t)$ on $L^2(G)$ decay exponentially proceeds by using
the Kunze-Stein convolution inequality, which asserts that $\left\|\lambda_G(\beta_t)\right\| \leq C_r \left\|\beta_t\right\|_{L^r(G)}$
for any $1 \leq r < 2$. Thus since any (coarsely) admissible family has exponential
volume growth, the exponential decay of the convolution norms, and hence the
operator norms, is established. In order to estimate the decay rate, we note that
(coarsely) admissible averages can be dominated by radial averages; namely, they
are (K, C)-radial for a fixed constant C and a fixed maximal compact subgroup K.
The convolution norm of a bi-K-invariant average is given simply by integration
against the Harish-Chandra Ξ-function, and thus an upper bound for $\left\|\pi_X^0(\beta_t)\right\|$ is
obtained.

4.3.1.2 The exponential strong maximal inequality

While the discussion related to the decay of the operator norms when π_X^0 is an L^p-representation applies to coarsely admissible families, the proof of the exponential strong maximal inequality makes essential use of the Hölder-regularity property of Hölder-admissible families, together with a rough-monotonicity property satisfied by the families, and the exponential decay of the operator norms. The method proceeds according to the following recipe, developed in [MNS] and [N4] and formulated in [N6].

1. Let $f \in L^2(X)$ be a nonnegative function on X. Then, since

$$\pi(\beta_t) f(x) \leq C\pi(\beta_{[t]+1}) f(x),$$

   we clearly have

$$\left\| \sup_{t>0} \pi(\beta_t) f(x) \right\|_{L^2(X)} \leq C \left\| \sup_{n \in \mathbb{N}} \pi(\beta_n) f(x) \right\|_{L^2(X)} \leq C' \|f\|_{L^2(X)}.$$

2. The previous argument extends to every L^p, $1 < p < \infty$, using the Riesz-Thorin interpolation theorem.

3. The estimate $\|\pi(\beta_t)\|_{L_0^2(X)} \leq C \exp(-\theta t)$ implies an exponential strong maximal inequality for any *sequence* of operators $\exp(\frac{1}{2}\theta t_k)\beta_{t_k}$ in L_0^2, where t_k is a sequence such that the sum of the norms converges. Repeat the argument in $L_0^p(X)$.

4. Now distribute $\lfloor \exp(\frac{1}{4}\theta n) \rfloor$ equally spaced points in the interval $[n, n+1]$. Then approximate $\pi(\beta_t)f$ by $\pi(\beta_{t_n})f$ using the closest point t_n to t in the sequence t_k. Estimate the difference using the exponential strong maximal inequality for the entire sequence β_{t_k}, and the local Hölder regularity of the family β_t, applied when f is a bounded function to the points t and t_k.

5. The previous argument gives an (L^∞, L^2) exponential strong maximal inequality, which says that the exponential-maximal function for $f \in L^\infty(X)$ has an L^2-norm bound in terms of the L^∞-norm of f. Now interpolate against the usual strong maximal inequality in L^p proved in the second step, using the analytic interpolation theorem.

Thus the proof of the exponential strong maximal inequality is complete, provided the representation π_X^0 has the property that a sufficiently high tensor power embeds in a multiple of the regular representation, or more generally that $\left\| \pi_X^0(\beta_t) \right\|$ decays exponentially.

4.3.2 Spectral gap, well balancedness, total weak-mixing

The existence of a strong spectral gap in a unitary representation of an S-algebraic group is equivalent to the existence of a spectral gap for all the restrictions of the representation to the simple components of the group. It is certainly possible that an action of a product group $G = G(1) \times G(2)$ will have a spectral gap but that the action restricted to one of the two components will fail to have a spectral gap. In this case, it is not true that there exists a sufficiently high tensor power of the representation π_X^0 which embeds in a multiple of the regular representation. As was the case with the notion of irreducible actions, we are not aware of a general criterion which will ensure that an action of G induced from an action of a lattice subgroup has a strong spectral gap. Since induced actions play a crucial role and serve to reduce the ergodic theorems for the lattice to the ergodic theorems for G, we must establish the ergodic theorems for G without assuming a strong spectral gap, otherwise the generality of the ergodic theorems for the lattice will be severely restricted. Since we are seeking quantitative estimates, we must introduce the necessary condition of well balancedness, which is a quantitative form of the balancedness condition. Given this condition, we use the unitary representation theory of S-algebraic groups and in Theorem 5.11 establish that indeed $\left\| \pi_X^0(\beta_t) \right\|$ decays exponentially. The arguments use a tensor power argument similar to the spectral transfer principle, radialization, and estimates of spherical functions. Having established the exponential decay of the norms, this allows us to use the method described in the previous section and completes the proof.

Finally, we note that a standing assumption, which we will use, is total weak-mixing, namely, the absence of any nontrivial irreducible finite-dimensional subspace invariant under any of the component subgroups of G. Exponential decay of the operator norms fails in certain finite-dimensional representations, as we noted in §3.8, so finite-dimensional representations must be excluded.

4.4 STATEMENT OF ERGODIC THEOREMS
FOR LATTICE SUBGROUPS

Theorems 4.1 and 4.2 will be used to derive corresponding results for *arbitrary* measure-preserving actions of lattice subgroups of G, provided that G does not admit nontrivial finite-dimensional unitary representation τ. This condition is necessary, and without it the formulation of ergodic theorems for the lattice must take into account the possible limiting values of $\tau(\lambda_t)$, as noted in Remark 4.4. Thus we will formulate our results for lattices Γ contained in G^+ since G^+ does have the desired property. When G is a Lie group, this amounts just to assuming the lattice is contained in the connnected component in the Hausdorff topology. In general, $\Gamma \cap G^+$ is a subgroup of finite index in Γ, since Γ is finitely generated and every finitely generated subgroup of G/G^+ is finite, but recall that for simplicity we assume $[G : G^+] < \infty$.

In the absence of a spectral gap, we will prove the following.

Theorem 4.5. *Let G be an S-algebraic group as in Definition 3.4 and let Γ be a lattice subgroup contained in G^+. Let G_t be an admissible one-parameter family (or an admissible sequence) in G^+ and $\Gamma_t = \Gamma \cap G_t$. Let (X, μ) be an arbitrary ergodic probability measure–preserving action of Γ. Then the averages λ_t satisfy the strong maximal inequality in (L^p, L^r) for $p \geq r \geq 1$, $(p, r) \neq (1, 1)$. If the* *action induced to G^+ is irreducible, the averages λ_t also satisfy*

1. *the mean ergodic theorem in L^p for $1 \le p < \infty$.*

2. *the pointwise ergodic theorem in L^p for $1 < p < \infty$, assuming in addition that G_t are left-radial.*

The same conclusions hold when the induced action is reducible, provided the family G_t is left-radial and balanced (for the mean theorem) or standard, well balanced, and boundary-regular (for the pointwise theorem).

In the presence of a spectral gap, we will prove the following.

Theorem 4.6. *Let G be an S-algebraic group as in Definition 3.4 and let Γ be a lattice contained in G^+. Let G_t be a Hölder-admissible one-parameter family (or an admissible sequence) in G^+ and $\Gamma_t = \Gamma \cap G_t$. Let (X, μ) be an arbitrary probability measure–preserving action of Γ. Assume either that the representation of G^+ induced by the representation of Γ on $L^2(X)$ has a strong spectral gap or that it has a spectral gap and the family G_t is well balanced. Then the averages λ_t satisfy the following:*

1. *the exponential mean ergodic theorem in (L^p, L^r) for $p \ge r \ge 1$, provided $(p, r) \ne (1, 1)$ and $(p, r) \ne (\infty, \infty)$.*

2. *the exponential strong maximal inequality in (L^p, L^r) for $p > r \ge 1$.*

3. *the exponential pointwise ergodic theorem in (L^p, L^r): for every $f \in L^p(X)$, $1 < p < \infty$, and almost every $x \in X$,*

$$\left| \pi_X(\lambda_t) f(x) - \int_X f \, d\mu \right| \le B_{p,r}(f, x) e^{-\zeta t},$$

where $\zeta = \zeta_{p,r} > 0$ and $r < p$.

Remark 4.7.

1. Regarding the assumptions of Theorem 4.5, we note that the action of G^+ induced from the Γ-action is indeed often (but perhaps not always) irreducible. In addition to the obvious case where G is simple, irreducibility holds (at least for groups over fields of zero characteristic) whenever the lattice is irreducible and the Γ-action is mixing [St, Cor. 3.8]. Another important case where it holds is when the lattice is irreducible and the Γ-action is via a dense embedding in a compact group [St, Thm. 2.1] or more generally when the Γ-action is isometric.

2. Regarding the assumptions of Theorem 4.6, we note that the unitary representation of G induced from the unitary representation of Γ on $L^2(X)$ always has a spectral gap provided the Γ-action on (X, μ) does, and it often (but perhaps not always) has a strong spectral gap. Indeed, by [M, Ch. III, Prop. 1.11], if the lcsc group G has a spectral gap in $L_0^2(G/\Gamma)$ and the Γ-representation on $L_0^2(X)$ has a spectral gap, then so does the representation

induced to G. The existence of a spectral gap in $L_0^2(G/\Gamma)$ has long been established for all lattices in S-algebraic groups. Thus, when the sets G_t are well balanced and left-radial, the conclusions of Theorem 4.6 hold, provided only that the action of Γ on (X, μ) has a spectral gap. If the induced action is irreducible and G has property T, then the induced representation has a strong spectral gap and any admissible family G_t will do.

Natural radial averages which satisfy all the required properties, and thus also the ergodic theorems, exist in abundance. To be concrete, let us concentrate on one family of examples and generalize Theorem 1.12 to the S-algebraic context.

Let G be an S-algebraic group as in Definition 3.4 and let ℓ denote the standard $CAT(0)$-metric on the symmetric space X or the Bruhat-Tits building X (or their product) associated to G. Let $G_t = \{g \in G \,;\, \ell(g \cdot o, o) \leq t\}$, where $o \in X$ is a fixed choice of origin, and let β_t be the associated averages. Let $\Gamma \subset G^+$ be a lattice subgroup, let $\Gamma_t = G_t \cap \Gamma$, and let λ_t be the uniform averages on Γ_t.

Theorem 4.8. *Let the notation be as in the preceding paragraph.*

1. *The averages β_t satisfy the mean, maximal, and pointwise ergodic theorems in every ergodic action of G (as in Theorem 4.1). If the action has a spectral gap, then β_t satisfy the exponential mean, maximal, and pointwise ergodic theorems (as in the conclusion of Theorem 4.2).*

2. *The averages λ_t satisfy the mean, maximal, and pointwise ergodic theorems in every ergodic action of Γ (as in Theorem 4.5). If the action has a spectral gap, then λ_t satisfy the exponential mean, maximal, and pointwise ergodic theorems (as in the conclusion of Theorem 4.6).*

3. *In every isometric action of Γ on a compact metric space, preserving an ergodic probability measure of full support, the averages λ_t become equidistributed (as in the conclusion of Theorem 6.14).*

Theorem 4.8 is a consequence of Theorems 3.18, 4.1, 4.2, 4.5, 4.6, and 6.14.

Of course, since the arguments in Chapter 7 employed to prove Theorem 3.18 apply whenever G_t satisfies certain growth and regularity conditions, Theorem 4.8 applies to more general families of averages.

4.5 ERGODIC THEOREMS FOR LATTICE SUBGROUPS: OVERVIEW

4.5.1 Induced actions and approximation

Our approach to proving the ergodic theorems for an action of a lattice subgroup $\Gamma \subset G$ on a space (X, μ) is based on the idea of inducing the Γ-action to G, thus obtaining an action of G on a space $Y = (G \times X)/\Gamma$, with an invariant probability measure $\nu = m_{G/\Gamma} \times \mu$. We then reduce the ergodic theorems for the averages λ_t supported on $\Gamma \cap G_t$ acting on $L^p(X)$ to the ergodic theorems for the averages β_t supported on $G_t \subset G$ and acting on $L^p(Y)$. This involves developing a series of approximation arguments, which become progressively more precise and more quantitative, culminating in the exponential pointwise ergodic theorems for lattice averages. The essence of the matter is to estimate the ergodic averages $\pi_X(\lambda_t)\phi(x)$ given by

$$\frac{1}{|\Gamma \cap G_t|} \sum_{\gamma \in \Gamma \cap G_t} \phi(\gamma^{-1}x), \quad \phi \in L^p(X),$$

above and below by the ergodic averages $\pi_Y(\beta_{t\pm C})F_\varepsilon(y)$, namely, by

$$\frac{1}{m_G(G_{t\pm C})} \int_{g\in G_{t\pm C}} F_\varepsilon(g^{-1}y)dm_G(g).$$

The link between the two expressions is given by setting $y = (h, x)\Gamma$ and

$$F_\varepsilon((h, x)\Gamma) = \sum_{\gamma \in \Gamma} \chi_\varepsilon(h\gamma)\phi(\gamma^{-1}x), \quad F \in L^p(Y),$$

where χ_ε is the normalized characteristic function of an identity neighborhood $\mathcal{O}_\varepsilon$. We assume that $\phi \geq 0$. Now the ergodic averages $\pi_Y(\beta_{t\pm C})F_\varepsilon$ can be rewritten in full as

$$\sum_{\gamma \in \Gamma} \left(\frac{1}{m_G(G_{t\pm C})} \int_{g\in G_{t\pm C}} \chi_\varepsilon(g^{-1}h\gamma) \right) \phi(\gamma^{-1}x),$$

and we would like the expression in parentheses to be equal to 1 when (say) $\gamma \in \Gamma \cap G_{t-C}$ and equal to 0 when (say) $\gamma \notin \Gamma \cap G_{t+C}$ in order to be able to compare it to $\pi_X(\lambda_t)\phi$. A favorable lower bound arises if $\chi_\varepsilon(g^{-1}h\gamma) \neq 0$ and $g \in G_{t-C}$ imply that $\gamma \in G_t$, and a favorable upper bound arises if for $\gamma \in \Gamma \cap G_t$ the support of $\chi_\varepsilon(g^{-1}h\gamma)$ is contained in G_{t+C}. Thus favorable lower and upper estimates depend only on the regularity properties of the sets G_t, specifically on the stability property under perturbations by elements h in a fixed neighborhood, and on the properties of $m_G(G_t)$.

Therefore, taking some fixed $\mathcal{O}_\varepsilon$ and C, the ordinary strong maximal inequality for averaging over λ_t follows from the ordinary strong maximal inequality for averaging over β_t. It follows that for lattice actions, as for actions of the group G, the maximal inequality holds in great generality and requires only coarse admissibility.

The mean ergodic theorem for λ_t requires a considerably sharper argument, and in particular requires passing to $\varepsilon \to 0$, namely, $m_G(\mathcal{O}_\varepsilon) \to 0$. The uniform quantitative estimate inherent in the definition of (Lipschitz or Hölder) admissibility is utilized and is matched against the the quantity $m_G(\mathcal{O}_\varepsilon)^{-1}$ which the approximation procedure introduces. Thus finiteness of the upper local dimension ϱ_0 (see Definition 3.5) associated with the family $\mathcal{O}_\varepsilon$ is essential.

The quantitative mean ergodic theorem requires in addition an effective estimate of the decay of the operator norms $\left\| \pi_Y^0(\beta_t) \right\|_{L_0^p(Y)}$. This estimate plays an indispensable role and allows the approximation argument to proceed, provided that the averages are Hölder-admissible. As a by-product of the proof, we obtain an effective exponential decay estimate of the norms $\left\| \pi_X^0(\lambda_t) \right\|_{L_0^p(X)}$ for $\varrho_0 < p < \infty$. An application of the Riesz-Thorin interpolation argument allows us to cover the full range $1 < p < \infty$.

4.5.2 Pointwise theorems and the invariance principle

Another fundamental point is the invariance principle, namely, the fact that for any given Borel function F on Y, the pointwise ergodic theorem for $\pi_Y(\beta_t)$ holds for a set of points which contain a strictly G-invariant conull set. Since the induced action $Y = (G \times X)/\Gamma$ is a G-equivariant bundle over G/Γ, this implies that for *every* point $y\Gamma$, the set of points $x \in X$ where the pointwise ergodic theorem holds is conull in X. This allows us to deduce that the set of points where $\pi_X(\lambda_t)\phi(x)$ converges is also conull in X.

To prove the exponential pointwise ergodic theorem for λ_t, we first use the method outlined in the case of G-actions. This method employs the operator norm decay result together with Hölder continuity and rough monotonicity and yields an exponential strong maximal inequality for the family λ_t. This inequality is then extended further to cover the full range $1 < p < \infty$ by using the analytic interpolation theorem, interpolating between the exponential strong maximal inequality for $p > \varrho_0$, and the ordinary strong maximal inequality for $p > 1$ (which was established earlier).

We note that establishing the main term of the lattice point–counting problem arises from the mean ergodic theorem for G-actions. Here we take X to the trivial action of Γ on a point, so that the induced action is the G-action on G/Γ. The error term arises from the quantitative mean ergodic theorem for G-actions, again for the same space. However, we treat these special cases separately, establishing them first, and then use the result in the respective mean ergodic theorem.

4.6 VOLUME REGULARITY AND VOLUME ASYMPTOTICS: OVERVIEW

As is already clear from the discussion so far, the regularity properties of the sets G_t are crucial in our approach. We recall that for the purpose of establishing the pointwise ergodic theorems for the averages β_t, the uniform local Lipschitz property of $\log m_G(G_t)$ is essential in the absence of a spectral gap, while the uniform local Hölder property is essential in its presence. As to the (Lipschitz or Hölder) stability property of the sets G_t under left and right perturbation by elements in $\mathcal{O}_\varepsilon$, it is essential for the purpose of deducing the ergodic theorems for λ_t from those of β_t.

Broadly speaking, since spectral methods play a very prominent role in our discussion, the stability and regularity properties of β_t become one of the main technical tools. They allow, among other things, the use of derivative arguments appearing in the square function estimate, the approximation of the operators β_t by a sufficiently dense *sequence* β_{t_k}, and the quantitative reduction of the analysis of the discrete averages λ_t on Γ to the absolutely continuous averages β_t on G. In effect, the regularity of β_t combined with spectral theory compensates for the failure of the classical large-scale methods of amenable ergodic theory associated with polynomial volume growth, covering argument, and the transfer principle.

We have thus been motivated to establish such regularity results in considerable generality, allowing for a diverse array of families G_t, arising from a variety of distance functions. This problem is quite demanding technically and requires the use of a broad range of techniques. In the case of connected semisimple Lie groups and norms or distances associated with Riemannian or affine symmetric spaces, our methods are motivated by and rely on those developed in [DRS], [EM], and [EMS]. The main tool is the existence of a polar coordinate decomposition of the form $G = KAH$ and a detailed study of its properties.

However, our results are formulated in the generality of S-algebraic groups, and this necessitates further discussion to handle product groups over Archimedean and non-Archimedean places. This discussion contains three main results pertaining to balls G_t on $G = G(1) \times \cdots \times G(N)$.

1. Theorem 3.15 considers balls G_t which are of the form

$$G_t = \left\{ g = (g_1, \ldots, g_N) ; \sum_{i=1}^{N} a_i d_i(g_i) < t \right\},$$

where either $d_i = \ell_i(g_i u_i, v_i)$, where ℓ_i is the standard $CAT(0)$-metrics on the symmetric space or the Bruhat-Tits building, or d_i is *any* linear norm in a faithful linear representation (over any local field). The structure theory of semisimple groups is employed to show that when G is connected, these averages are admissible. When there are also totally disconnected factors present, further convolution arguments are employed to show that the resulting averages are still admissible. Typically these averages are not balanced, but it is always possible to choose the positive constants a_i so that the resulting averages are balanced.

2. Theorem 3.18 considers balls G_t which are of the form

$$G_t = \left\{ g = (g_1, \ldots, g_N) ; \sum_{i=1}^{N} \ell_i(u_i, g_i u_i)^p < t^p \right\},$$

where $1 < p < \infty$ and ℓ_i are as above, so that these averages are all radial, namely, bi-K-invariant where K is a suitable maximal compact subgroup. Detailed information on the Cartan polar coordinates decomposition and the radial volume density is then used, together with further convolution arguments, to deduce that the associated averages are well balanced and boundary-regular. These two properties are crucial for the arguments appearing in the proof of some of the ergodic theorems. In addition, when at least one Archimedean factor is present, the family G_t is shown to be admissible.

3. Theorem 3.16 considers the balls defined by any standard height function on a semisimple algebraic group and establishes that β_t are Hölder-admissible. The first step in the proof is to consider any real algebraic affine variety X with a regular volume form, $\Psi : X \to \mathbb{R}$ a regular proper function, and $G_t = \{\Psi < t\}$. Hironaka resolution of singularities is applied together with Tauberian arguments in order to

(i) deduce that the function $t \mapsto \mathrm{vol}(\{\Psi < t\})$ is uniformly locally Hölder-continuous,

(ii) compute its asymptotics with an error term.

We note that the idea of using Hironaka resolution of singularities to establish meromorphic continuation goes back to the works of Atiyah [A] and Bernstein and Gelfand [BeGe]. In the second step, to handle the case of balls defined by a height function on an arbitrary semisimple S-algebraic group, Denef's theorem on the rationality of the height integrals is employed in the finite places, and finally convolution arguments are employed to handle the case of a general product.

Chapter Five

Proof of ergodic theorems for S-algebraic groups

In the present chapter we will prove the ergodic theorems for admissible or Hölder-admissible averages on an S-algebraic group G stated in Theorems 4.1 and 4.2. We will distinguish two cases, namely, whether the action of G on X has a spectral gap or not. As noted already in Chapter 4, the arguments that will be employed below in these two cases are quite different, but both use spectral theory in a material way. Therefore we will begin by recalling the relevant facts from spectral theory. In order to consider all S-algebraic groups, it is convenient to work in the generality of groups admitting an Iwasawa decomposition, which we define below. This setup will have the added advantage that it incorporates a large class of subgroups of groups of automorphism of products of Bruhat-Tits buildings. This class contains more than all semisimple algebraic groups and S-algebraic groups and is of considerable interest. We will then proceed to the proofs of the ergodic theorems, first in the presence of a spectral gap and then in its absence. Finally, we will also state and prove the invariance principle for actions of G.

5.1 IWASAWA GROUPS AND SPECTRAL ESTIMATES

5.1.1 Iwasawa groups

We begin by introducing the class of groups to be considered.

Definition 5.1. Groups with an Iwasawa decomposition.

1. An lcsc group G has an *Iwasawa decomposition* if it has two closed amenable subgroups K and P, with K compact and $G = KP$.

2. The *Harish-Chandra Ξ-function* associated with the Iwasawa decomposition $G = KP$ of the unimodular group G is given by

$$\Xi(g) = \int_K \delta^{-1/2}(gk)dk,$$

where δ is the left modular function of P extended to a left-K-invariant function on $G = KP$. (Thus if m_P is a left Haar measure on P, $\delta(p)m_P$ is right P-invariant, and a Haar measure on G is given by $dm_G = dm_K \delta(p)dm_P$).

Convention. The definition of an Iwawasa group involves a choice of a compact subgroup and an amenable subgroup. When G is the group of F-rational points of a semisimple algebraic group defined over a locally compact nondiscrete field F, G

admits an Iwasawa decomposition, and below we can and will always choose K to be a good maximal compact subgroup and P a corresponding minimal F-parabolic group (see [T] or [Si]). This choice will be naturally extended in the obvious way to S-algebraic groups.

5.1.2 Spectral estimates

Since Iwasawa groups possess a compact subgroup admitting an amenable complement, it is natural to consider the decomposition of a representation of G to K-isotypic subspaces. In general, let G be an lcsc group, K be a compact subgroup, and $\pi : G \to \mathcal{U}(\mathcal{H})$ be a strongly continuous unitary representation, where $\mathcal{U}(\mathcal{H})$ is the unitary group of the Hilbert space $\mathcal{H}$.

Definition 5.2. K-finite vectors and strongly L^p representations.

1. A vector $v \in \mathcal{H}$ is called K-finite (or $\pi(K)$-finite) if its orbit under $\pi(K)$ spans a finite-dimensional space.

2. The unitary representation σ of G is *weakly contained* in the unitary representation π if for every $F \in L^1(G)$ the estimate $\|\sigma(F)\| \leq \|\pi(F)\|$ holds. Clearly, if σ is strongly contained in π (namely, equivalent to a subrepresentation), then it is weakly contained in π.

3. The representation π is called a *strongly L^p representation* if there exists a dense subspace $\mathcal{J} \subset \mathcal{H}$ such that the matrix coefficients $\langle \pi(g)v, w \rangle$ belong to $L^p(G)$ for $v, w \in \mathcal{J}$.

We recall the following spectral estimates, which will play an important role below.

Theorem 5.3. Tensor powers and norm estimates. *For any lcsc group G and any strongly continuous unitary representation π,*

1. *[CHH, Thm. 1] If π is strongly $L^{2+\varepsilon}$ for all $\varepsilon > 0$, then π is weakly contained in the regular representation λ_G.*

2. *[C2] and [Ho] (see [HT] for a simple proof) If π is strongly L^p and n is an integer satisfying $n \geq p/2$, then $\pi^{\otimes n}$ is strongly contained in $\infty \cdot \lambda_G$.*

3. *[N4, Thm 1.1, Prop. 3.7] If π is strongly L^p and n_e is an even integer satisfying $n_e \geq p/2$, then $\|\pi(\mu)\| \leq \|\lambda_G(\mu)\|^{1/n_e}$ for every probability measure μ on G. If the probability measures μ and μ' satisfy $\mu \leq C\mu'$ as measures on G, then $\|\pi(\mu)\| \leq C' \|\lambda_G(\mu')\|^{1/n_e}$.*

We begin by stating the following basic spectral estimates for Iwasawa groups, which are straightforward generalizations of [CHH].

Theorem 5.4. *Let $G = KP$ be a unimodular lcsc group with an Iwasawa decomposition and π a strongly continuous unitary representation of G. Let v and w be two K-finite vectors and denote the dimensions of their spans under K by d_v and d_w. Then the following estimates hold, where Ξ is the Harish-Chandra Ξ-function.*

1. *If π is weakly contained in the regular representation, then*

$$|\langle \pi(g)v, w \rangle| \leq \sqrt{d_v d_w} \, \|v\| \, \|w\| \, \Xi(g).$$

2. *If π is strongly $L^{2k+\varepsilon}$ for all $\varepsilon > 0$, then*

$$|\langle \pi(g)v, w \rangle| \leq \sqrt{d_v d_w} \, \|v\| \, \|w\| \, \Xi(g)^{1/k}.$$

Proof. Part 1 is stated in [CHH, Thm. 2] for semisimple algebraic groups, but the same proof applies for any unimodular Iwasawa group.

Part 2 is stated in [CHH] for irreducible representations of semisimple algebraic groups, but the same proof applies to an arbitrary representation of unimodular Iwasawa groups since it reduces to part 1 after taking a k-fold tensor product. $\square$

Remark 5.5.

1. The quality of the estimate in Theorem 5.4 depends of course on the structure of G. For example, if P is normal in G (so that G is itself amenable), then P is unimodular if G is. Then $\delta(g) = 1$ for $g \in G$, and the estimate is trivial.

2. In the other direction, Theorem 5.4 will be most useful when the Harish-Chandra function is indeed in some $L^p(G)$, $p < \infty$, so that Theorem 5.3 applies.

3. For semisimple algebraic groups, the Ξ-function is in fact in $L^{2+\varepsilon}$ for all $\varepsilon > 0$, which is a well-known result due to Harish-Chandra ([HC1], [HC2], and [HC3]).

4. When Ξ is in some L^q, $q < \infty$, Theorem 5.4(1) implies that any representation with a tensor power which is weakly contained in the regular representation is strongly L^p for some p. This assertion of course also uses the density of K-finite vectors, which is a consequence of the Peter-Weyl theorem.

We will use Theorem 5.3(3) to bound the operator norm of a given measure by that of its radialization. Thus in particular, for a Haar-uniform probability measure on a (K, C)-radial set, the operator norm in an L^p-representation is bounded in terms of the convolution norm of the Haar-uniform probability measure on its radialization.

We now state the following result, which summarizes a number of results from [C2], [HM], and [BW] in a form convenient for our purposes.

Theorem 5.6. L^p-representations ([C2], [HM], and [BW]). *Let F be a locally compact nondiscrete field. Let G denote the F-rational point of an algebraically connected semisimple algebraic group which is almost F-simple. Let π be a unitary representation of G without nontrivial finite-dimensional G-invariant subspaces (or equivalently, without G^+-invariant unit vectors).*

1. *If the F-rank of G is at least two, then π is strongly $L^{p+\varepsilon}$ (for all $\varepsilon > 0$), with $p = p(G) < \infty$ depending only on G.*

2. *If the F-rank of G is one, then any unitary representation π admitting a spectral gap (equivalently, which does not contain an asymptotically invariant sequence of unit vectors) is strongly L^p for some $p = p(\pi) < \infty$. In particular, every irreducible infinite-dimensional representation has this property.*

3. *When G is an S-algebraic group as in Definition 3.4 and π is such that every simple component has a spectral gap and no nontrivial finite-dimensional subrepresentations, then π is an L^p-representation.*

Proof. When G is simple and π is irreducible, parts 1 and 2 are stated in [C2, Thms. 2.4.2 and 2.5.2] in the Archimedean case, and in [HM, Thm. 5.6] in general (based on the theory of leading exponents in [BW]). It follows that if $n \geq p/2$, then the tensor power of any n such representations can be embedded in a multiple of the regular representation of G (see, e.g., [HT]). For a general unitary representation π of G with a spectral gap, a direct integral argument shows that $\pi^{\otimes n}$ embeds in a multiple of the regular representation, and hence by Theorem 5.4, K-finite matrix coefficients of π are indeed in $L^p(G)$.

Since every irreducible unitary representation of an S-algebraic group G is a tensor product of irreducible unitary representations of the simple components, part 3 follows from a direct integral argument and from parts 1 and 2. $\qquad\square$

We remark that an explicit estimate of the best exponent $p = p_G$ when G is simple and has split rank at least two is given by [L] and [LZ] in the Archimedean case, and by [O] for p-adic groups.

5.2 ERGODIC THEOREMS IN THE PRESENCE OF A SPECTRAL GAP

5.2.1 Spectral gap and exponential strong maximal inequality

Consider now an ergodic measure-preserving action of an lcsc group G on a probability space (X, μ) and the associated unitary representation π_X^0 on $L_0^2(X)$. When the representation admits a spectral gap, it can be established in certain situations that the operator norms $\left\| \pi_X^0(\beta_t) \right\|$ decay exponentially (say) in t. In that case, as noted in Chapter 4, an exponential strong maximal inequality holds for general Hölder families of probability measures ν_t on G. The precise formulation follows, and for a proof we refer the reader to [N4] and to [MNS], where the relation between the rate of exponential decay and the parameters p and r below is fully explicated.

Theorem 5.7. Exponential strong maximal inequality [N4, Thm. 4]. *Let G be an lcsc group and assume that the family of probability measures ν_t is uniformly locally Hölder-continuous in the total variation norm, namely, $\|\nu_{t+\varepsilon} - \nu_t\| \leq C\varepsilon^a$ for all $t \geq t_0$ and $0 < \varepsilon \leq \varepsilon_0$. Assume also that it is roughly monotone, namely, $\nu_t \leq C\nu_{[t]+1}$, where C is fixed.*

1. Assume that $\pi_X^0(\nu_t)$ have exponentially decaying norms in $L_0^2(X)$. Then the exponential strong maximal inequality in (L^p, L^r) holds in any probability measure–preserving action of G, and thus also the exponential pointwise ergodic theorem holds in (L^p, L^r), for any $p > r > 1$.

2. In particular, if the representation π_X^0 on $L_0^2(X)$ is a strongly L^p representation and ν_t have exponentially decaying norms as convolution operators on $L^2(G)$, then the previous conclusion holds.

Remark 5.8. For future reference, let us recall the following simple observation from [N4]. For *any sequence* of averages ν_n, the exponential strong maximal inequality in (L^p, L^p), $1 < p < \infty$, is an immediate consequence of the exponential decay condition $\left\| \pi_X^0(\nu_t) \right\| \leq Ce^{-n\theta}$. So is the exponential pointwise ergodic theorem, and both follow simply by considering the bounded operator $\sum_{n=0}^{\infty} e^{n\theta/2} \pi_X^0(\nu_n)$ on $L_0^2(X)$ and then using Riesz-Thorin interpolation.

It is clear that Hölder-admissible families are roughly monotone and satisfy the Hölder continuity condition stated in the assumptions of Theorem 5.7. Consequently, in order to prove Theorem 4.2, it suffices to establish the exponential decay of the operator norms when the averages are Hölder-admissible. We now turn to estimating operator norms, a task we will in fact be able to accomplish for much more general averages.

5.2.2 Bounds on operator norms in L^p-representations

5.2.2.1 Roughly radial sets

Note that according to Theorem 5.3(3), it is possible to bound the operator norm of (say) a given Haar-uniform probablity measure on a (K, C)-radial set in terms of the convolution norm of its radialization. We formulate this fact as follows.

Proposition 5.9. Radialization estimate. *Let* $G = KP$ *be an lcsc unimodular Iwasawa group and* π *a strongly* L^p-*representation. Let* B *be a set of positive finite measure and* $\beta = \chi_B / \mathrm{vol}(B)$.

1. *If* B *is bi-K-invariant, then*

$$\|\lambda(\beta)\| = \frac{1}{m(B)} \int_B \Xi_G(g) dm_G(g).$$

2. *If* B *is a* (K, C)-*radial set and* $\tilde{B} = KBK$, *then*

$$\|\pi(\beta)\| \leq C' \left(\frac{\int_{KBK} \Xi(g) dm_G(g)}{m_G(KBK)} \right)^{1/n_e} = C' \left\| \lambda(\tilde{\beta}) \right\|^{1/n_e},$$

provided n_e *is even and* $n_e \geq p/2$.

Both statments hold of course for every absolutely continuous (K, C)-*radial probability measure (with the obvious definition of a* (K, C)-*radial measure).*

Proof. 1. Let us first recall the following estimate from [CHH], dual to the C. Herz majorization principle. The spectral norm of the convolution operator $\lambda_G(F)$ on $L^2(G)$ is estimated by

$$\|\lambda_G(F)\| \leq \int_G \left(\int_K \int_K |F(kgk')|^2 \, dk dk' \right)^{1/2} \Xi(g) dg$$

for any measurable function F on G for which the right hand side is finite. Clearly, when $F = \chi_B$ is bi-K-invariant, we obtain the upper bound

$$\|\lambda_G(\beta)\| \leq \frac{1}{m(B)} \int_B \Xi(g) dg.$$

The equality in part 1 for bi-K-invariant sets B follows from the fact that since P is amenable, the representation of G on G/P induced from the trivial representation of P is weakly contained in the regular representation of G. Here Ξ is a diagonal matrix coefficient of the representation in $L^2(G/P)$ by definition, so that

$$\frac{1}{m(B)} \int_B \Xi_G(g) dm_G(g) \leq \|\lambda_G(\beta)\|.$$

2. Now consider a set B which is (K, C)-radial. Clearly, a comparison of the convolutions with the normalized probability measures β and $\tilde{\beta}$ immediately gives

$$\|\lambda_G(\beta)\| \leq \frac{C}{m_G(KBK)} \|\lambda_G(\chi_{KBK})\|.$$

Applying the previous inequality to the bi-K-invariant measure $\tilde{\beta}$ uniformly distributed on KBK, we clearly obtain

$$\|\lambda_G(\beta)\| \leq \frac{C}{m_G(KBK)} \int_{KBK} \Xi(g) dm_G(g).$$

Now by the definition of weak containment, the same inequality also holds for unitary representations π which are weakly contained in the regular representation. Taking tensor powers and using Theorem 5.3, we obtain a norm bound for $\|\pi(\beta)\|$ in L^p-representations as well. $\square$

5.2.2.2 *Bounding norms using the Kunze-Stein phenomenon*

To complement our discussion, let us recall, as noted in [N4, Thm. 3,4], that it is also possible to estimate the operator norm of general (not necessarily (K, C)-radial) averages using the Kunze-Stein phenomenon, provided the representation is strongly L^p, $p < \infty$.

1. Indeed, for the spectral norm, namely, when $f \in C_c(G)$ and $\beta = \chi_B/m_G(B)$ (B a bounded set), we have by the Kunze-Stein phenomenon, provided $1 < p < 2$,

$$\|\beta * f\|_{L^2(G)} \leq C_p \|\beta\|_{L^p(G)} \|f\|_{L^2(G)}.$$

Taking $p = 2 - \varepsilon$ and f ranging over functions of unit $L^2(G)$-norm, we conclude that

$$\|\lambda(\beta)\| \leq C_\varepsilon' \|\beta\|_{L^{2-\varepsilon}(G)} = C_\varepsilon' m_G(B)^{-(1-\varepsilon)/(2-\varepsilon)} \leq C_\varepsilon'' m_G(B)^{-1/2+\varepsilon}.$$

2. If the representation π satisfies $\pi^{\otimes n_e} \subset \infty \cdot \lambda_G$ (e.g., if it is strongly L^p and n_e is even with $n_e \geq p/2$), then by part 1 and Theorem 5.3(3) (see also [N4, Thm. 4]), $\|\pi(\beta)\| \leq C_\varepsilon m_G(B)^{-1/(2n_e)+\varepsilon}$.

3. It follows from part 2, that if β_t is a family of Haar-uniform averages on the sets G_t and $\pi^{\otimes n_e} \subset \infty \cdot \lambda_G$, then the rate of volume growth of G_t can be used to give a lower bound the largest parameter θ satisfying $\|\pi(\beta_t)\| \leq C_\varepsilon e^{(-\theta+\varepsilon)t}$ (for every $\varepsilon > 0$) as follows:

$$\theta = \liminf_{t \to \infty} -\frac{1}{t} \log \|\pi(\beta_t)\| \geq \frac{1}{2n_e} \liminf_{t \to \infty} \frac{1}{t} \log m_G(B_t).$$

Remark 5.10. Let us note the remarkable fact that this favorable norm estimate holds for every family G_t whose volume grows exponentially. Given that the representation is L^p, the estimate requires no regularity or stability conditions of any kind.

5.2.3 Bounding operator norms in nonintegrable representations

We now turn to establishing the necessary decay estimates for the norms of the operators $\pi(\beta_t)$, in the case of a general representation, which may not satisfy the L^p-integrability condition. In this case, the condition of well balancedness becomes crucial. We will prove the following.

Theorem 5.11. Exponential decay of operator norms. *Let G be S-algebraic as in Definition 3.4 and let σ be a totally weak-mixing unitary representation of G. Let G_t be a coarsely admissible one-parameter family or sequence. Assume that σ has a strong spectral gap or that it has a spectral gap and G_t are well balanced. Then for some C and $\delta = \delta_\sigma > 0$ depending on σ and G_t,*

$$\|\sigma(\beta_t)\| \leq C e^{-\delta t}.$$

Proof. 1. *The case of L^p-representations.* We have already shown in Proposition 3.22 that coarsely admissible families are (K, C)-radial, and so let us use Proposition 5.9. If the representation σ is strongly L^p, then denoting the radializations by $\widetilde{G}_t = K G_t K$, the norm of $\sigma(\beta_t)$ is estimated by

$$\|\sigma(\beta_t)\| \leq C' \left(\frac{\int_{\widetilde{G}_t} \Xi_G(g) dm_G(g)}{m_G(\widetilde{G}_t)} \right)^{1/n_e},$$

where $n_e \geq p/2$ is even and Ξ_G is the Harish-Chandra Ξ-function.

Recall that coarsely admissible sets satisfy the growth condition $S^n \subset G_{an+b}$. It follows of course that their radializations satisfy $\widetilde{S}^n \subset \widetilde{G}_{a'n+b'}$ for a compact bi-K-invariant generating set $\widetilde{S}$. The standard estimates of $\Xi_G(g)$ (see [HC1], [HC2], and [HC3]) now imply that $\|\sigma(\beta_t)\|$ decays exponentially.

Now the assumption that the representation is strongly L^p is satisfied when the representation is totally weak-mixing and has a strong spectral gap, as noted in Theorem 5.6(3). Indeed, (almost) every irreducible representation appearing in the direct integral decomposition of σ w.r.t. a simple subgroup must be infinite-dimensional and has a fixed tensor power contained in a multiple of the regular representation.

We note that the strong spectral gap assumption is necessary here. Indeed, let σ be the tensor product of an irreducible principal series representation of $G = \mathrm{PSL}_2(\mathbb{Q}_p)$ and a weak-mixing representation of G which admits an asymptotically invariant sequence of unit vectors. Then σ has a spectral gap (as a representation of $G \times G$) and is totally weak-mixing but is not strongly L^p for any finite p.

2. *The case of nonintegrable representations.* To handle the case where σ is totally weak-mixing but is not strongly L^p, let us recall that we now assume G_t are well balanced and also coarsely admissible. It follows that the radialized sets $\tilde{G}_t$ are also well balanced. Indeed, since $G_t \subset K G_t K \subset G_{t+c}$ for every coarsely admissible family, clearly also (referring to Definition 3.17) $G_t \cap G(J)Q \subset KG_t K \cap G(J)Q \subset G_{t+c} \cap G(J)Q$ for every compact subset Q of $G(I)$. Taking $t = an$ and $Q = S_I^n$ and using the fact that G_t are well balanced, the claim follows. Now, if $\bar{\sigma}$ is the representation conjugate to σ, then for any probability measure ν on G and every vector u, we have

$$\|\sigma(\nu)u\|^2 = \langle \sigma(\nu^* * \nu)u, u \rangle \leq \int_G |\langle \sigma(g)u, u \rangle| \, d(\nu^* * \nu)(g)$$

$$\leq \left(\int_G |\langle \sigma(g)u, u \rangle|^2 \, d(\nu^* * \nu) \right)^{1/2}$$

$$= \left(\langle \sigma \otimes \bar{\sigma}(\nu^* * \nu)(u \otimes \bar{u}), u \otimes \bar{u} \rangle \right)^{1/2} \leq \| \sigma \otimes \bar{\sigma}(\nu)(u \otimes \bar{u}) \|.$$

Hence it suffices to prove that $\| \sigma \otimes \bar{\sigma}(\beta_t) \|$ decays exponentially. But note that the diagonal matrix coefficients of $\sigma \otimes \bar{\sigma}$ which we are now considering are all nonnegative. Thus for each vector u and every (K, C)-radial measure ν, using Jensen's inequality and the previous argument,

$$\|\sigma(\nu)u\|^4 = \langle \sigma(\nu^* * \nu)u, u \rangle^2 \leq \langle \sigma \otimes \bar{\sigma}(\nu^* * \nu)(u \otimes \bar{u}), u \otimes \bar{u} \rangle$$

$$\leq C^2 \langle \sigma \otimes \bar{\sigma}(\tilde{\nu}^* * \tilde{\nu})(u \otimes \bar{u}), u \otimes \bar{u} \rangle = C^2 \| \sigma \otimes \bar{\sigma}(\tilde{\nu})(u \otimes \bar{u}) \|^2.$$

We conclude that if the norm of $\sigma \otimes \bar{\sigma}(\tilde{\beta}_t)$ decays exponentially, so does the norm of $\sigma(\beta_t)$. Now if σ has a spectral gap and is totally weak-mixing, then $\sigma \otimes \bar{\sigma}$ has the same properties. This claim follows from Theorems 5.6 and 5.4. Indeed, if an asymptotically invariant sequence of unit vectors exists in $\sigma \otimes \bar{\sigma}$, then there is also such a sequence which consists of K-invariant vectors, so that we can restrict our attention to the spherical spectrum. But a sequence consisting of convex sums of products of normalized positive-definite spherical functions cannot converge to 1 uniformly on compact sets unless the individual spherical functions that occur in them have the same property, and this contradicts the spectral gap assumption on σ. Total weak-mixing follows from a direct integral decomposition and the fact that the tensor product of two irreducible infinite-dimensional unitary representations of a simple group do not have finite-dimensional subrepresentations.

Thus we are reduced to establishing the norm decay of a bi-K-invariant coarsely admissible well-balanced family in a totally weak-mixing unitary representation with a spectral gap, which we continue to denote by σ.

To estimate the norm of $\sigma(\tilde{\beta}_t)$ under these conditions, write $G = G(1) \times G(2)$, where $G(1)$ has property T and $G(2)$ is a product of groups of split rank one. Any irreducible unitary representation of G is a tensor product of irreducible unitary representations of $G(1)$ and $G(2)$. Since $\tilde{\beta}_t$ are bi-K-invariant measure, we can clearly restrict our attention to infinite-dimensional spherical representations, namely, those containing a K-invariant unit vector, and then estimate the matrix coefficient given by the spherical function. The nonconstant spherical functions $\varphi_s(g)$ on G are given by $\varphi_{s_1}(g_1)\varphi_{s_2}(g_2)$, where at least one of the factors is nonconstant. If $\varphi_{s_1}(g_1)$ is nonconstant, then again it is a product of spherical functions on the simple components of $G(1)$, one of which is nonconstant. The function $\varphi_{s_1}(g_1)$ is then bounded, according to Theorems 5.4 and 5.6, by the function $\Phi : g_1 \mapsto \Xi_{G(1)/H}(p_{G(1)/H}(g_1))^{1/n}$ for some fixed n depending only on $G(1)$, where $G(1)/H$ is one of the simple factors of $G(1)$ and $p_{G(1)/H}$ is the projection onto it. Using the standard estimate of the Ξ-function, together with our assumption that the G_t are well balanced, we conclude that for $\eta_1 > 0$ depending only on $G(1)$ and G_t,

$$\frac{1}{m_G(G_t)} \int_{G_t} |\varphi_s(g)|\, dm_G(g) \le Ce^{-\eta_1 t}. \tag{5.1}$$

Now, if $\varphi_{s_1}(g_1)$ is constant, then the representation of G in question is trivial on $G(1)$ and factors to a nontrivial irreducible representation of $G(2)$. The spherical functions of the complementary series on a split rank one group are the only ones we need to consider, and they have a simple parametrization as a subset of an interval. We can take the interval to be say $[0, 1]$, where 0 corresponds to the Harish-Chandra function and 1 to the constant function. The function $\varphi_{s_2}(g_2)$ is a product of spherical functions on the real rank factors. The assumption that the original representation $\sigma = \pi_X^0$ has a spectral gap implies that for some $\delta > 0$, the parameter of the spherical function of at least one factor is less than $1 - \delta$. The bound $\delta > 0$ depends only on the original representation σ and is uniform over the representations of $G(2)$ that occur. It is then easily seen, using the estimate of the Ξ-function in the split rank one case and the fact that $\tilde{G}_t$ are well balanced, that $\varphi_s(g)$ also satisfies the estimate (5.1) for some $\eta_1' > 0$. This gives a uniform bound over all irreducible unitary spherical representations of G that are weakly contained in σ, and it follows that $\left\| \pi_X^0(\tilde{\beta}_t) \right\| \le Ce^{-\eta t}$, where $\eta = \eta(\sigma) > 0$. $\qquad \square$

Proof of Theorem 4.2. As already noted, the rough monotonicity and uniform local Hölder continuity property that appear in the assumptions of Theorem 5.7 are clearly satisfied by one-parameter Hölder-admissible families or admissible sequences.

Furthermore, a coarsely admissible family satisfies the exponential decay condition in $L_0^2(X)$, provided the latter is an L^p-representation, or if the family is well balanced, in view of Theorem 5.11. Thus the exponential mean, pointwise, and maximal ergodic theorems hold for Hölder-admissible one-parameter families.

Finally, the case of a sequence of admissible averages does not require an appeal to Theorem 5.7, just to the remark following it, as well as to Theorem 5.11.

This concludes the proof of all parts of Theorem 4.2. $\qquad \square$

Remark 5.12. Note that the Hölder admissibility assumption is sufficient also for the proof of parts 4 and 5 of Theorem 6.3 and part 4 of Theorem 6.4.

5.3 ERGODIC THEOREMS IN THE ABSENCE OF A SPECTRAL GAP, I

We now turn to the proof of Theorem 4.1 and consider actions which do not necessarily admit a spectral gap. We start by proving the strong maximal inequality for admissible families (and some more general ones), and in the sections that follows we establish pointwise convergence (to the ergodic mean) on a dense subspace. In the course of that discussion the mean ergodic theorem, namely, norm convergence (to the ergodic mean) on a dense subspace, will be apparent. As noted in Chapter 4, these three ingredients suffice to prove Theorem 4.1 completely (see, e.g., [N6] for a full discussion).

5.3.1 The strong maximal inequality

Consider a family β_t of probability measures on G, where β_t is the Haar-uniform average on G_t and G_t are coarsely admissible. As noted in Proposition 3.22, coarse admissibility implies that G_t is a family of (K, C)-radial sets with C fixed and independent of t. As before, we denote by $\tilde{\beta}_t$ the Haar uniform averages of the sets $\tilde{G}_t = KG_tK$ and again note that $\beta_t \leq C\tilde{\beta}_t$ as measures on G. Hence for $f \geq 0$ we have almost surely

$$f_\beta^*(x) = \sup_{t>t_0} \pi_X(\beta_t)f(x) \leq C \sup_{t>t_0} \pi_X(\tilde{\beta}_t)f(x) = Cf_{\tilde{\beta}}^*(x),$$

so that it suffices to prove the strong maximal inequality for the averages $\tilde{\beta}_t$. Furthermore, if G_t is coarsely admissible, then clearly so are the sets $\tilde{G}_t = KG_tK$.

Recall now that that coarse admissibility implies $(\mathcal{O}_r, D)$-ampleness for some constants r and D, as noted in Proposition 3.22. Ampleness is the key to the strong maximal inequality, which actually holds here in great generality, as noted in Chapter 4. The strong maximal inequality for $\tilde{\beta}_t$ follows from the following.

Theorem 5.13. Strong maximal inequality for ample sets (see [N5, Thm. 3]). *Let G be an S-algebraic group as in Definition 3.4 and let K be of finite index in a maximal compact subgroup. Let (X, μ) be a totally weak-mixing probability measure–preserving action of G. For a set $E \subset G$ of positive finite measure, let ν_E denote the Haar-uniform average supported on E. Fix positive constants $r > 0$ and $D > 1$ and consider the maximal operator*

$$\mathcal{A}^*f(x) = \sup\{|\pi_X(\nu_E)f(x)| \ : \ E \subset G \text{ and } E \text{ is } (\mathcal{O}_r, D, K)\text{-ample}\}.$$

Then $\mathcal{A}^*f$ satisfies the maximal inequality ($1 < p \leq \infty$):

$$\|\mathcal{A}^*f\|_{L^p(X)} \leq B_p(G, r, D, K)\|f\|_{L^p(X)}.$$

Proof. Theorem 5.13 is proved in full for connected semisimple Lie groups with finite center in [N5]. The same proof applies to our present more general context

without essential changes, as we now briefly note. First, if G is a totally disconnected almost simple algebraically connected noncompact algebraic group with
property T, the analog of [N5, Thm. 2], namely, the exponential strong maximal
inequality for the cube averages defined there follows from the exponential decay
of the norm of the *sequence* (in this case) of cube averages. The exponential decay
in the space orthogonal to the invariants is assured by our assumption that the action
is totally weak-mixing, which implies that Theorems 5.6 and 5.4 can be applied.
Then standard estimates of the Ξ-function yield the desired conclusion for the cube
averages.

Second, for groups of split rank one the strong maximal inequality for the sphere
averages is established in [NS1] when the Bruhat-Tits tree has even valency, and the
same method gives the general case (using the description of the spherical function
given, e.g., in [N1]). The fact that the strong maximal inequality for cube averages
holds on product groups if it holds for the components is completely elementary, as
in [N5, §2]. Finally, the fact that the strong maximal inequality for cube averages
implies the strong maximal inequality for an ample set in a given group depends
only on analysis of the volume density associated with the Cartan decomposition
and thus only on the root system, and the argument in [N5, §4] generalizes without
difficulty.

This establishes the strong maximal inequality for every S-algebraic group as in
Definition 3.4.   □

5.4 ERGODIC THEOREMS IN THE ABSENCE
OF A SPECTRAL GAP, II

In the present section we begin the proof of the existence of a dense subspace
of functions in $L^2(X)$ where $\pi_X^0(\beta_t)f(x)$ converges almost surely, using spectral theory and a Sobolev-type argument. We first discuss the latter, and we will
make crucial use of the absolute continuity property established for admissible
one-parameter of families averages in Proposition 3.13. This property implies that
$t \mapsto \beta_t$ is almost surely differentiable (in t and w.r.t. the $L^1(G)$-norm) with a
globally bounded derivative.

When β_t are the Haar-uniform averages on an increasing family of compact sets
G_t, almost sure differentiabilty of $t \mapsto \beta_t$ in the $L^1(G)$-norm is equivalent to the
almost sure existence of the limit

$$\lim_{\varepsilon \to 0} \frac{1}{\varepsilon} \frac{m_G(G_{t+\varepsilon}) - m_G(G_t)}{m_G(G_t)},$$

and for admissible families, the limit is uniformly bounded as a function of t. The
almost sure differentiability allows us to make use of a Sobolev-type argument
developed originally in [N2].

The uniform local Lipshitz continuity for the averages is a stronger property
than uniform local Hölder continuity, which was the underlying condition in the
case where the action has a spectral gap. However, as noted in Chapter 4, the
Lipshitz condition allows us to dispense with the assumption of exponential decay

of $\left\| \pi_X^0(\beta_t) \right\|$.

Note also that by Theorem 5.13 the (ordinary) strong L^p-maximal inequality holds for much more general averages, namely, under the sole conditions that G_t are (K, C)-radial and their radializations are $(\mathcal{O}_r, D)$-ample averages on an S-algebraic group. It is only pointwise convergence on a dense subspace that requires the additional regularity assumption of almost sure differentiability, which follows from the uniform local Lipshitz condition.

Finally, we remark that the argument we give below is based solely on the spectral estimates described in the previous sections. Thus it does not require extensive considerations related to classification of unitary representations and applies to all semisimple algebraic groups (and other Iwasawa groups).

Let (G, m_G) denote an lcsc group G with a left Haar measure m_G. Let $G_t, t > 0$, be an admissible family. Then (3.5) and (3.6) are satisfied for $t \geq t_0 > 0$. Without loss of generality, we may assume that $m_G(G_{t_0}) > 0$ and $G_t = \cap_{s > t} G_s, t \geq t_0$. Note that an admissible family necessarily satisfies $\cup_{t \geq t_0} G_t = G$. Let $g \mapsto |g|$ be the gauge $|g| = \inf\{s \geq t_0 ; g \in G_s\}$. The intersection condition implies that G_t are determined by their gauge via $G_t = \{g \in G ; |g| \leq t\}$. Thus the gauge is a measurable proper function with values in $[t_0, \infty)$. Define $\beta_t, t \in \mathbb{R}_+$, to be the one-parameter family of probability measures with compact supports on G, absolutely continuous w.r.t. Haar measure, whose density is given by the function $\frac{1}{m_G(G_t)} \chi_{G_t}(g)$. The map $t \mapsto \beta_t$ is a uniformly locally Lipshitz function from $[t_0, \infty)$ to $L^1(G)$, w.r.t. the norm topology, by assumption.

We let $S_t = \{g ; |g| = t\}$, and clearly $G_t = G_{t_0} \coprod_{t_0 < s \leq t} S_s$ is a disjoint union. The map $g \mapsto |g|$ projects Haar measure on G onto a measure on $[t_0, \infty)$, which is an absolutely continuous measure w.r.t. Lebesgue measure on (t_0, ∞), by Proposition 3.13. The measure disintegration formula gives the representation $m_G = m_G|_{G_{t_0}} + \int_{t_0}^{\infty} m_r dr$, where m_r is a measure on S_r defined for almost all $r > t_0$. Thus we can write for any $F \in C_c(G)$ and $t \geq t_0$,

$$
\begin{aligned}
\beta_t(F) &= \frac{\int_{G_t} F dm_G}{m_G(G_t)} = \frac{1}{m_G(G_t)} \left(\int_{G_{t_0}} F dm_G + \int_{G_t \backslash G_{t_0}} F dm_G \right) \\
&= \frac{1}{m_G(G_t)} \left(\int_{G_{t_0}} F dm_G + \int_{t_0}^{t} \frac{m_r(F)}{m_r(S_r)} m_r(S_r) dr \right) \\
&= \frac{m_G(G_{t_0})}{m_G(G_t)} \beta_{t_0}(F) + \int_{t_0}^{t} \partial \beta_r(F) \psi_t(r) dr.
\end{aligned}
$$

Here $\partial \beta_r = m_r / m_r(S_r)$ is a probability measure on S_r (for almost every r), and the density $\psi_t(r)$ is given by $\psi_t(r) = m_r(S_r)/m_G(G_t)$. Clearly, $\psi_t(r)$ is a measurable function, defined almost surely w.r.t. Lebesgue measure on $[t_0, \infty)$, and is almost surely positive for $t_0 \leq r \leq t$.

For any given continuous function $F \in C_c(G)$, we claim that $\beta_t(F)$ is an absolutely continuous function on (t_0, ∞). Note that $\beta_t(F)$ is a sum of two terms, the first given by an almost surely differentiable function with bounded derivative. Indeed, when a uniform local Lipschitz condition is satisfied by $\log m_G(G_t)$, namely, when $m_G(G_{t+\varepsilon}) \leq (1 + c\varepsilon) m_G(G_t)$ for $0 < \varepsilon \leq \varepsilon_0$, and all $t \geq t_0$, it results in a uniform estimate of the ratio of the "area of the sphere" (i.e., S_t) to the "volume of the ball" (i.e., G_t). Namely, by Lebesgue's differentiation theorem on the real line, we have the following.

Lemma 5.14. *Assume that $G_t, t > 0$ is an admissible one-parameter family. Then for $t \geq t_0$,*

$$\frac{m_G(G_{t+\varepsilon}) - m_G(G_t)}{m_G(G_t)} = \frac{\int_t^{t+\varepsilon} m_r(S_r)dr}{m_G(G_t)} \leq c\varepsilon,$$

so that for almost all $t > t_0$ we have the uniform bound

$$\frac{m_t(S_t)}{m_G(G_t)} = \frac{1}{m_G(G_t)} \lim_{\varepsilon \to 0} \frac{1}{\varepsilon} \int_t^{t+\varepsilon} m_r(S_r)dr \leq c.$$

In particular, $\frac{d}{dt}\left(\frac{1}{m_G(G_t)}\right)$ exists almost surely on (t_0, ∞) and is a bounded function.

The second term in $\beta_t(F)$ is given by integration against the L^1-density $\partial\beta_r(F)\psi_t(r)$ (which is almost everywhere defined). We can therefore conclude that $\beta_t(F)$ is differentiable almost everywhere, and its derivative is given as follows.

Proposition 5.15. *Assume that the family G_t, $t > t_0$, gives rise to an absolutely continuous measure on (t_0, ∞), as above. Then for almost all $t > t_0$,*

$$\frac{d}{dt}(\beta_t(F)) = \frac{m_t(S_t)}{m_G(G_t)} (\partial\beta_t - \beta_t)(F).$$

Proof. We compute

$$\frac{d}{dt}\beta_t = \frac{d}{dt}\left(\frac{1}{m_G(G_t)}\left[m_G(G_{t_0})\beta_{t_0} + \int_{t_0}^t m_r dr\right]\right)$$

$$= \left(\frac{1}{m_G(G_t)}\right)'\left[m_G(G_{t_0})\beta_{t_0} + \int_{t_0}^t m_r dr\right] + \frac{1}{m_G(G_t)} \cdot m_t$$

$$= -\frac{m_G(G_t)'}{m_G(G_t)^2}\left[m_G(G_{t_0})\beta_{t_0} + \int_{t_0}^t m_r dr\right] + \frac{m_t(S_t)}{m_G(G_t)}\partial\beta_t$$

$$= \frac{m_t(S_t)}{m_G(G_t)}(\partial\beta_t - \beta_t).$$

We have used $\frac{d}{dt}\int_{t_0}^t m_r dr = m_t$ and $m_G(G_t)' = m_t(S_t)$ for almost all $t > t_0$, which are both consequences of the Lebesgue differentiation theorem on the real line. □

Since $\beta_t(F)$ is absolutely continuous, it is given by integration against an L^1-density equal almost everywhere to its derivative. It follows, in particular, that for every $F \in C(G)$ and for *every* $t > s > t_0$, the following identity holds:

$$\beta_t(F) - \beta_s(F) = \int_s^t \frac{d}{dr}\beta_r(F)dr.$$

This identity being valid for *every* $F \in C_c(G)$, it follows that the corresponding equality holds between the underlying (signed) Borel measures on G, namely,

$$\beta_t - \beta_s = \int_s^t \frac{d}{dr}\beta_r dr = \int_s^t \frac{m_r(S_r)}{m_G(G_r)} (\partial \beta_r - \beta_r) dr.$$

Let us now use the fact that the derivative $\frac{d}{dr}\beta_r$ is a multiple of the difference between two Borel probability measures on G (for almost every $r \in (t_0, \infty)$). Every bounded Borel measure on G naturally gives rise to a bounded operator on the representation space. We can thus conclude the following relations between the corresponding operators defined in any G-action.

Corollary 5.16. Differentiable vectors.

1. *For any strongly continuous unitary representation and any vector $u \in \mathcal{H}$, $t \mapsto \pi(\beta_t)u$ is almost surely differentiable in $t \in (t_0, \infty)$ (strongly, namely, in the norm topology), and the following holds:*

$$\pi(\beta_t)u - \pi(\beta_s)u = \int_s^t \frac{d}{dr}\pi(\beta_r)u dr.$$

2. *Consider a measurable G-action on a standard Borel probability space and a function $u(x) \in L^p(X)$, $1 \le p < \infty$, for which $g \mapsto u(g^{-1}x)$ is continuous in g for almost every $x \in X$. Then the expression $\pi(\beta_t)u(x) = \int_G u(g^{-1}x)d\beta_t(g)$ is differentiable in t for almost every $x \in X$ and almost every $t \in (t_0, \infty)$, and the following almost sure identity holds:*

$$\pi(\beta_t)u(x) - \pi(\beta_s)u(x) = \int_s^t \frac{d}{dr}\pi(\beta_r)u(x) dr.$$

Remark 5.17. Note that in Corollary 5.16(2), the space of vectors u satisfying the assumptions is norm-dense in the corresponding Banach space. Indeed, the subspace contains all convolutions of the form $C_c(G) * L^\infty(X)$, which is clearly norm-dense in $L^p(X)$.

Our spectral approach uses direct integral decomposition for the representation of G in $L^2(X)$, and we thus assume that G is a group of type I. As is well known, this assumption is satisfied by all S-algebraic groups. We note further that it is typically the case for an Iwasawa group $G = KP$ that K is large in G. Namely, in every *irreducible* representation π of G, the space of (K, τ)-isotypic vectors in $\mathcal{H}_\pi$ is finite-dimensional for every irreducible representation τ of K. In particular, this property does indeed hold for every S-algebraic group.

5.5 ERGODIC THEOREMS IN THE ABSENCE OF A SPECTRAL GAP, III

We can now state the following convergence theorem for admissible families of averages.

Theorem 5.18. Pointwise convergence on a dense subspace for admissible families. *Let G be an lcsc group of type I and K a compact subgroup. Let β_t be an admissible family of averages on G.*

Consider a K-finite vector u in the τ-isotypic component under K in an irreducible infinite-dimensional unitary representation ρ of G (we do not assume τ is irreducible). Assume that for every such ρ there exist $\delta = \delta_{\rho,\tau} > 0$ and a positive constant $C_\rho(\tau, \delta)$ (both independent of u) such that

$$\int_{t_0}^{\infty} e^{\delta r} \left(\|\rho(\beta_r)u\| + \|\rho(\partial\beta_r)u\| \right)^2 dr \leq C_\rho(\tau, \delta) \|u\|^2 .$$

Then in any measure-preserving weak-mixing action of G on (X, μ), there exist closed subspaces $\mathcal{H}(\tau, \delta, N) \subset L_0^2(X)$ where $\pi_X^0(\beta_t) f(x) \to 0$ almost surely for $f \in \mathcal{H}(\tau, \delta, N)$. The convergence is of course also in the L^2-norm. Furthermore, the union

$$\bigcup_{\delta>0, N<\infty, \tau\in\widehat{K}} \mathcal{H}(\tau, \delta, N)$$

is dense in $L_0^2(X)$.

Proof. Our proof of Theorem 5.18 is divided into two parts, as follows.

1. *Direct integrals.* Any unitary representation π is of the form

$$\pi = \int_{z\in\Sigma_\pi}^{\oplus} \rho_z dE(z),$$

where $\Sigma_\pi \subset \widehat{G}$ is the spectrum of the representation and E is the corresponding (projection-valued) measure. Furthermore, the Hilbert space of the representation admits a direct integral decomposition $\mathcal{H}_\pi = \int_{z\in\Sigma_\pi}^{\oplus} \mathcal{H}_z dE(z)$. In particular, any vector $u \in \mathcal{H}$ can be identified with a measurable section of the family $\{\mathcal{H}_z ; z \in \Sigma_\pi\}$, namely, $u = \int_{z\in\Sigma_\pi}^{\oplus} u_z dE(z)$, where $u_z \in \mathcal{H}_z$ for E–almost all $z \in \Sigma_\pi$. Clearly, u belongs to the τ-isotypic component of π if and only if u_z belongs to the τ-isotypic of ρ_z for E–almost all z. To see this, note that u is characterized by the equation $u = \pi(\chi_\tau)u$, where χ_τ is the character of the representation τ on K, namely, as being in the range of a self-adjoint projection operator. The projection operator commutes with all the spectral projections since the latter commute with all the unitary operators $\pi(g)$, $g \in G$, and hence also with their linear combinations. This implies $\rho_z(\chi_\tau)u_z = u_z$ E–almost surely. Given another vector v in the τ-isotypic component, the following spectral representation is valid.

$$\langle \pi(g)u, v \rangle = \int_{z\in\Sigma_\pi} \langle \rho_z(g)u_z, v_z \rangle \, dE_{u,v}(z),$$

where $E_{u,v}$ is the associated (scalar) spectral measure. Thus the K-finite vectors of π (with variance τ under K) are integrals (w.r.t. the spectral measure), of K-finite vectors with the same variance, associated with irreducible unitary representations ρ_z of G.

2. *Sobolev space argument.* Our second step is a Sobolev space argument following [N2, §7.1]. We assume the estimate stated in Theorem 5.18 for every irreducible nontrivial representation ρ. Given $\tau \in \widehat{K}$, for each ρ, we define

$$\delta_{\rho,\tau} = \frac{1}{2} \sup \left\{ \delta > 0 \; ; \; \sup_{\|u\|=1} \int_{t_0}^{\infty} e^{\delta r} \left(\|\rho(\beta_r)u\| + \|\rho(\partial\beta_r)u\|\right)^2 dr < \infty \right\},$$

where u ranges over the τ-isotypic component.

By our assumption, $\delta_{\rho,\tau}$ is positive and finite, and in addition, this function of ρ is measurable w.r.t. the spectral measure. Therefore, for this choice of $\delta = \delta_{\rho,\tau}$, the estimator (for u ranging over the τ-isotypic component)

$$C_\rho(\tau, \delta_\rho) = 2 \sup_{\|u\|=1} \int_{t_0}^{\infty} e^{r\delta_{\rho,\tau}} \left(\|\rho(\beta_r)u\| + \|\rho(\partial\beta_r)u\|\right)^2 dr$$

is also measurable w.r.t. the spectral measure. Therefore we can consider the measurable sets

$$A(\tau, \delta, N) = \left\{ z \in \Sigma_\pi \; ; \; \delta_{\rho_z,\tau} > \delta, \; C_{\rho_z}(\tau, \delta_{\rho_z,\tau}) \leq N \right\}$$

and the corresponding closed spectral subspaces

$$\mathcal{H}(\tau, \delta, N) = \int_{A(\tau,\delta,N)}^{\oplus} \mathcal{H}_z dE(z).$$

Thus, in particular, in these subspaces the decay of τ-isotypic K-finite matrix coefficients is exponentially fast, with at least a fixed positive rate determined by δ.

Now note that the subspace of differentiable vectors in $A(\tau, \delta, N)$ which are invariant under the projection $\pi(\chi_\tau)$ is norm-dense in the τ-isotypic subspace. Indeed, the subspace $\pi(\chi_\tau * C_c(G))(A(\tau, \delta, N))$ consists of differentiable τ-isotypic vectors and is dense in the τ-isotypic subspace. Furthermore, given a differentiable vector u (of this form, namely, a convolution) in the τ-isotypic subspace, u_z is also differentiable, for E–almost all $z \in \Sigma_\pi$ (w.r.t. the spectral measure), since for $f \in C_c(G)$,

$$\pi(\chi_\tau * f * \chi_\tau)u = \int_{z \in A(\tau,\delta,N)} \rho_z(\chi_\tau * f * \chi_\tau)u_z dE(z).$$

We can now use the first part of Corollary 5.16, together with standard spectral theory, and conclude that for a differentiable vector $u \in A(\tau, \delta, N)$, the following spectral identity holds for every $t > s > t_0$ and for every $u, v \in L^2(X)$:

$$\langle (\pi(\beta_t) - \pi(\beta_s)) u, v \rangle = \int_{z \in \Sigma_\pi} \langle (\rho_z(\beta_t) - \rho_z(\beta_s)) u_z, v_z \rangle \, dE_{u,v}(z)$$

$$= \int_{z \in \Sigma_\pi} \int_s^t \left\langle \frac{d}{dr}\rho_z(\beta_r)u_z, v_z \right\rangle dr \, dE_{u,v}(z).$$

Now using Corollary 5.16(2) and the fact that v above is allowed to range over $L^2(X)$, for each t and s, we have the following equality of functions in $L^2(X)$, namely, for almost all $x \in X$,

$$\pi(\beta_t)u(x) - \pi(\beta_s)u(x) = \int_s^t \frac{d}{dr}\pi(\beta_r)u(x)dr,$$

so that for any $t > s \geq M$, for almost all $x \in X$,

$$|\pi(\beta_t)u(x) - \pi(\beta_s)u(x)| \leq \int_M^\infty \left|\frac{d}{dr}\pi(\beta_r)u(x)\right| dr.$$

The averages β_t form an continuous family in the $L^1(G)$-norm, consisting of absolutely continuous measures on G, and the function $t \mapsto \pi(\beta_t)u(x)$ (where u is a convolution) is therefore a continuous function of t for almost every $x \in X$. Restricting our attention to these points x, we conclude that for all $M > 0$ and almost every x,

$$\limsup_{t,s\to\infty} |\pi(\beta_t)u(x) - \pi(\beta_s)u(x)| \leq \int_M^\infty \left|\frac{d}{dr}\pi(\beta_r)u(x)\right| dr,$$

and thus the set

$$\left\{ x ; \limsup_{t,s\to\infty} |\pi(\beta_t)u(x) - \pi(\beta_s)u(x)| > \zeta \right\}$$

is contained in the set

$$\left\{ x ; \int_M^\infty \left|\frac{d}{dr}\pi(\beta_r)u(x)\right| dr > \zeta \right\}.$$

We estimate the measure of the latter set by integrating over X and using the Cauchy-Schwartz inequality. We obtain, for any $\zeta > 0$ and $M > 0$, the following estimate:

$$\mu \left\{ x ; \int_M^\infty \left|\frac{d}{dr}\pi(\beta_r)u(x)\right| dr > \zeta \right\}$$

$$\leq \frac{1}{\zeta} \int_X \left(\int_M^\infty \left|\frac{d}{dr}\pi(\beta_r)u(x)\right| dr \right) d\mu(x)$$

$$\leq \frac{1}{\zeta} \int_M^\infty \left\|\frac{d}{dr}\pi(\beta_r)u\right\|_{L^2(X)} dr$$

$$\leq \frac{\exp(-M\delta/4)}{\zeta} \int_M^\infty e^{-r\delta/4}e^{r\delta/2}\left\|\frac{d}{dr}\pi(\beta_r)u\right\|_{L^2(X)} dr$$

$$\leq \frac{2\exp(-M\delta/2)}{\zeta\sqrt{M\delta/2}} \left(\int_M^\infty e^{r\delta}\left\|\frac{d}{dr}\pi(\beta_r)u\right\|_{L^2(X)}^2 dr \right)^{1/2}.$$

Using Proposition 5.15, it suffices to show the finiteness of the expression

$$\int_M^\infty e^{r\delta}\left\|\frac{m_r(S_r)}{m_G(G_r)}\pi(\partial\beta_r - \beta_r)u\right\|_{L^2(X)}^2 dr.$$

Using our assumption that the K-finite vector u has its spectral support in the set $A(\tau, \delta, N)$, we can write the last expression as

$$\int_M^\infty e^{\delta r} \int_{z \in A(\tau, \delta, N)} \left\| \frac{m_r(S_r)}{m_G(G_r)} \rho_z(\beta_r - \partial\beta_r)u_z \right\|^2_{\mathcal{H}_z} dE_{u,u}(z)dr.$$

Using the uniform bound given in Corollary 5.14, we can estimate by

$$\int_M^\infty e^{\delta r} \int_{z \in A(\tau, \delta, N)} c^2 \left(\|\rho_z(\beta_r)u_z\| + \|\rho_z(\partial\beta_r)u_z\| \right)^2 dE_{u,u}(z)dr.$$

Thus by definition of the space $A(\tau, \delta, N)$ and the fact that u is spectrally supported in this subspace, we conclude that the last expression is bounded by $c^2 N \|u\|^2 < \infty$.

We have established that $\pi_X^0(\beta_t)u(x)$ converges almost surely (exponentially fast) to the ergodic mean, namely, to zero. The fact that

$$\bigcup_{\delta>0, N<\infty, \tau \in \widehat{K}} \mathcal{H}(\tau, \delta, N)$$

is dense in $L_0^2(X)$ is a standard fact in spectral theory. This concludes the proof of Theorem 5.18. $\qquad\square$

To complete the proof of Theorem 4.1, we now need to verify that the assumptions of Theorem 5.18 are satisfied by S-algebraic groups. We begin with the following.

Theorem 5.19. *Let $G = G(1) \cdots G(N)$ be an S-algebraic group as in Definition 3.4. Let G_r be any family of bounded Borel sets and β_r the Haar-uniform probability measures. Let $\rho = \rho_1 \otimes \cdots \otimes \rho_N$, where each ρ_i is an irreducible unitary representation of $G(i)$ without $G(i)^+$-invariant unit vectors. Then there exist $\delta = \delta_\rho > 0$ and a constant C_1 (depending only on G and the family G_r) such that for every τ-isotypic vector u,*

1. *When G_r is a coarsely admissible one-parameter family (or sequence), we have*
$$\|\rho(\beta_r)u\| \le C_1(\dim \tau)e^{-\delta r} \|u\|.$$

2. *When G_r is a left-K-radial admissible family, with K a good maximal compact subgroup, for almost every r, we have*
$$\|\rho(\partial\beta_r)u\| \le C_1(\dim \tau)e^{-\delta r} \|u\|.$$

3. *In particular, when G_r is left-K-radial and admissible, there exists a constant $C_\rho(\tau, \delta) < \infty$ such that*
$$\int_{t_0}^\infty e^{\delta r} \left(\|\rho(\beta_r)u\| + \|\rho(\partial\beta_r)u\| \right)^2 dr \le C_\rho(\tau, \delta) \|u\|^2.$$

Proof.

1. By Theorem 5.6, ρ is strongly L^p for some $p = p(\rho) < \infty$, and then if $n \geq p/2$, $\rho^{\otimes n} \subset \infty \cdot \lambda_G$. Assume without loss of generality that n is even and then (see [N4, Thm. 1.1]), using Jensen's inequality, we obtain

$$\|\rho(\beta_r)u\|^{2n} \leq \left(\int_G |\langle \rho(g)u, u \rangle| \, d(\beta_r^* * \beta_r) \right)^n$$

$$\leq \int_G |\langle \rho(g)u, u \rangle|^n \, d(\beta_r^* * \beta_r)$$

$$= \int_G |\langle \rho^{\otimes n}(g)u^{\otimes n}, u^{\otimes n} \rangle| \, d(\beta_r^* * \beta_r)$$

$$\leq \dim(\tau)^n \|u\|^{2n} \int_G \Xi(g)d(\beta_r^* * \beta_r),$$

where we have used the estimate given in Theorem 5.4(1) for K-finite matrix coefficients in representations (weakly) contained in the regular representation. Now Ξ is nonnegative, and G_r are assumed coarsely admissible and hence (K, C)-radial. Thus we can multiply the last estimate by C^2 and then replace β_r by their radializations $\tilde{\beta}_r$.

Since we are considering S-algebraic groups, we can assume without loss of generality that K is a good maximal compact subgroup so that (G, K) is a Gelfand pair. Then Ξ defines a $*$-homomorphism of the commutative convolution algebra of bi-K-invariant functions $L^1(K \backslash G/K)$. We can therefore conclude that

$$\|\rho(\beta_r)u\|^{2n} \leq C^2(\dim \tau)^n \|u\|^{2n} \left(\int_G \Xi(g)d\tilde{\beta}_r(g) \right)^2.$$

Coarse admissibility implies the following growth property for G_r and their radializations $\tilde{G}_r$, namely, $S^n \subset G_{an+b}$ for a compact generating set S. Thus the desired result follows from the standard estimates of the Ξ-function of an S-algebraic group, which shows that the integral of Ξ_G on G_r decays exponentially in r.

2. Now consider the case of $\partial \beta_r$, which is a singular measure on G, supported on the "sphere" S_r. Arguing as in part 1, we still have

$$\|\rho(\partial \beta_r)u\|^{2n} \leq \dim(\tau)^n \|u\|^{2n} \int_G \Xi(g)d(\partial \beta_r^* * \partial \beta_r).$$

Now Ξ is bi-K-invariant for a good maximal compact subgroup K, and G_r (and thus β_r) are assumed to be left-K-invariant. It follows that $m_K * \partial \beta_r = \partial \beta_r$ for almost every r. Since Ξ_G is also a $*$-homomorphism of the bi-K-invariant measure algebra, we obtain

$$\int_G \Xi(g)d(\partial \beta_r^* * \partial \beta_r) = \int_G m_K * \Xi * m_K(g)d(\partial \beta_r^* * m_K * \partial \beta_r)$$

$$= \int_G \Xi_G(g)d(m_K * \partial \beta^* * m_K * \partial \beta_r * m_K)$$

$$= \left(\int_G \Xi(g)d\partial \beta_r \right)^2.$$

But $\partial \beta_r = m_r/m_r(S_r)$ is a probability measure supported on S_r, and clearly the growth property for G_r implies that S_r is contained in the complement of $S^{a_1[r]}$ for some $a_1 > 0$. Therefore the standard estimates of the Ξ-function again yield the desired result.

3. The last part is an immediate consequence of the previous two.   □

Completion of the proof of Theorem 4.1. The last step in the proof of Theorem 4.1 is to consider the various alternatives stated in its assumptions.

If the action is irreducible and totally weak-mixing, then almost every irreducible unitary representation ρ_z of G appearing in the direct integral decomposition of π_X^0 is indeed strongly L^p for some finite p. This follows from Theorem 5.6 since ρ_z is then a tensor product of infinite-dimensional irreducible representations of the simple constituent groups. In that case parts 1 and 2 of Theorem 5.19 apply, and the proof of the mean and the pointwise ergodic theorems for left-radial admissible one-parameter families in irreducible actions is complete, also taking into account that the strong maximal inequality is covered in all cases by Theorem 5.13.

Note that in the irreducible case, we can still apply part 1 of Theorem 5.19 to a coarsely admissible sequence, and this immediately yields the mean ergodic theorem and pointwise convergence almost surely on the dense subspace of vectors appearing there. Again using Theorem 5.13, this completes the proof of the mean and pointwise ergodic theorems for coarsely admissible sequences in irreducible actions.

Otherwise the action may be reducible, and we seek to prove the mean theorem when the left-radial averages are balanced, and the pointwise theorem when they are standard and well balanced. In the present case, each ρ_z is a tensor product of infinite-dimensional irreducible representations of some of the simple subgroups, and the trivial representations of the others. We can repeat the argument used in the second part of the proof of Theorem 5.11 and establish that $\|\rho_z(\beta_t)u\| \to 0$ using the assumption that β_t and hence $\tilde{\beta}_t$ are balanced, and that $\|\rho_z(\beta_t)u\| \to 0$ exponentially fast when β_t and hence $\tilde{\beta}_t$ are well balanced. Indeed, instead of integrating against (a power of) $\Xi_G(g)$ as above, we will now be integrating against (a power of) the Ξ-function lifted from some simple factor group. By the balanced or well-balanced assumption, the standard estimates of the Ξ-function yield the desired norm decay conclusion.

The last argument required to complete the proof of the pointwise theorem is the estimate of $\|\pi_X^0(\partial\beta_t)\|$ when the averages are standard, well balanced, and boundary-regular. In this case each defining distance ℓ (or d) on a factor group L obeys the estimate provided by Theorem 3.18(2), namely,

$$m_t(\partial G_t \cap \mathrm{proj}^{-1}(L_{\alpha t})) \leq Ce^{-\beta t}m_t(\partial G_t).$$

The total measure $m_t(S_t)$ on $\partial G_t \subset G$ is obtained as an iterated integral over the factor groups. Integrating against a power of the Ξ-function lifted from a factor group and using the decay of the Ξ-function, the required estimate follows.

This concludes the proof of all parts of Theorem 4.1 (and of course also Theorem 1.4).   □

Remark 5.20.

1. We note that our analysis applies to a general almost surely differentiable family of averages β_t (absolutely continuous w.r.t. Haar measure) and not only to those arising from Haar-uniform averages on admissible sets G_t as in Theorem 5.18.

2. We need only assume that the irreducible representations of G giving rise to the spectral decomposition of $L^2(X)$ satisfy the spectral estimates we have employed and not necessarily all representations of G. This is useful when considering the homogeneous space $X = G/\Gamma$, where G is a semisimple Adele group; see [GN].

Remark 5.21. Singular averages. An important problem that arises naturally in the present context is to extend the foregoing analysis to averages which are singular w.r.t. Haar measure. An obvious first step would be to establish a pointwise ergodic theorem for the family of "spherical averages" supported on the boundaries ∂G_t of the sets G_t, generalizing [N3], [N4], [NS2], and [CN]. However, to prove such results it is necessary to establish estimates of the *derivatives* of the τ-spherical functions. While the matrix coefficients themselves obey uniform decay estimates which are independent of the representation (provided, say, that it is L^p; see Theorems 5.3 and 5.4), this is no longer the case for their derivatives. For example, consider the principal series representations $\mathrm{Ind}_{MAN}^G 1 \otimes i\eta$ induced from a unitary character of A and the trivial representation of MN. These representations have matrix coefficients whose derivatives exhibit explicit dependence on the character η parametrizing the representation. Consequently, sufficiently sharp *derivative* estimates of matrix coefficients are inextricably tied up with classification, or at least parametrization, of the irreducible unitary representations of the group (see [CN] for more on this point).

We have avoided appealing to classification theory and refrained from establishing such derivative estimates in the present book. Instead we have utilized the fact that restricting to *Haar-uniform averages* on admissible sets, the distribution $\frac{d}{dt}\beta_t$ is a signed measure, so that we need only use the estimates of the spherical functions themselves (discussed in §5.1 and 5.2) in order to control it.

5.6 THE INVARIANCE PRINCIPLE AND STABILITY OF ADMISSIBLE AVERAGES

5.6.1 The set of convergence

It will be essential in our arguments below that for a family of admissible averages, the set where pointwise convergence of $\pi(\beta_t)f(x)$ holds contains a G-invariant set for each fixed measurable function f. We now establish this fact.

Let G be a locally compact second countable group with left Haar measure m_G. Consider as usual a measure-preserving action of G on a standard Borel space

$(X, \mathcal{B}, \nu)$, and for Borel subsets $G_t \subset G$, consider the averages

$$\beta_t = \frac{1}{m_G(G_t)} \int_{G_t} \delta_h \, dm_G(h).$$

Let us formulate the following invariance principle which applies to all quasi-uniform families (see Definition 3.20). This result generalizes [BR], where the case of ball averages on $SO(n, 1)$ was considered.

Theorem 5.22. *Let G be an lcsc group and suppose that $\{G_t\}_{t>0}$ is a quasi-uniform family, with β_t satisfying the pointwise ergodic theorem in $L^p(\nu)$. Let f be an (everywhere-defined) Borel-measurable function with $f \in L^p(\nu)$. Then there exists a G-invariant Lebesgue-measurable set $\Omega(f)$ of full measure such that for every $x \in \Omega(f)$,*

$$\lim_{t \to \infty} \pi(\beta_t) f(x) = \int_X f \, d\nu.$$

In particular, this holds for Hölder-admissible one-parameter families and admissible sequences on S-algebraic groups.

Proof. We can assume of course that f is real, and writing $f = f^+ - f^-$ for $f^+, f^- \in L^p(\nu)$, $f^+, f^- \geq 0$, and assuming that the theorem holds for f^+ and f^-, we can take

$$\Omega(f) = \Omega(f^+) \cap \Omega(f^-).$$

Hence, without loss of generality, we may assume that $f \geq 0$.

Consider then the conull measurable set of convergence

$$C = \left\{ x \in X : \lim_{t \to \infty} \pi(\beta_t) f(x) = \int_X f \, d\nu \right\}.$$

Note that C does in principle depend on the choice of f in its $L^p(\nu)$-class. Choose a countable dense set $\{g_i\}_{i \geq 1} \subset G$ and let

$$\Omega = \bigcap_{i \geq 1} g_i C.$$

Then Ω is a measurable set of full measure, and for every $x \in \Omega$ and every g_i, we have $g_i^{-1} x \in C$. Let $\delta > 0$ and take $\varepsilon > 0$ and $\mathcal{O}$ as in (3.10) and (3.11). We may also assume that $\mathcal{O}$ is symmetric. Then for any $g \in G$, there exists g_i such that $g_i \in g\mathcal{O}$. Hence, for sufficiently large t,

$$g_i G_{t-\varepsilon} \subset gG_t \subset g_i G_{t+\varepsilon}.$$

Therefore, for every $x \in X$, by nonnegativity of f,

$$\pi(\beta_t) f(g^{-1}x) = \frac{1}{m_G(G_t)} \int_{G_t} f(h^{-1}g^{-1}x) dm_G(h)$$

$$= \frac{1}{m_G(G_t)} \int_{gG_t} f(u^{-1}x) dm_G(u)$$

$$\leq \frac{1}{m_G(G_t)} \int_{g_i G_{t+\varepsilon}} f(u^{-1}x) dm_G(u)$$

$$\leq \frac{1+\delta}{m_G(g_i G_{t+\varepsilon})} \int_{g_i G_{t+\varepsilon}} f(u^{-1}x) dm_G(u)$$

$$= (1+\delta)\pi(\beta_{t+\varepsilon}) f(g_i^{-1}x).$$

This implies that for every $g \in G$ and $x \in \Omega$, since $g_i^{-1} x \in C$,

$$\limsup_{t \to \infty} \pi(\beta_t) f(g^{-1}x) \leq (1 + \delta) \int_X f \, d\mu$$

for every $\delta > 0$. Similarly, we show that

$$\liminf_{t \to \infty} \pi(\beta_t) f(g^{-1}x) \geq (1 + \delta)^{-1} \int_X f \, d\mu.$$

Thus for $g \in G$ and $x \in \Omega$, $g^{-1}x$ belongs to the convergence set C, and therefore let us take $\Omega(f) = G \cdot \Omega \subset C$. Then $\Omega(f)$ is strictly invariant under G, namely, $g\Omega(f) = \Omega(f)$ for every $g \in G$, and the complement of $\Omega(f)$ is a null set. Thus $\Omega(f)$ is a strictly invariant measurable set in the Lebesgue σ-algebra, namely, in the completion of the standard Borel structure on X with respect to the measure μ. □

An immediate corollary of the foregoing considerations is the following.

Corollary 5.23. *Let G be an lcsc group and suppose that $\{G_t\}_{t>0}$ is a quasi-uniform family, with β_t satisfying the pointwise ergodic theorem in $L^p(v)$. Then the Haar-uniform averages on gG_th also satisfy it for any fixed $g, h \in G$. In particular, this holds for Hölder-admissible one-parameter families and admissible sequences on S-algebraic groups.*

5.6.2 Stability of admissible averages under translations

When is the family gG_th itself already admissible if G_t is? This was asserted in Definition 1.1 for connected Lie groups. In this subsection we establish that in general the property of admissibility is stable under two-sided translations. Indeed, the sets $\mathcal{O}_\varepsilon$ we used to defined admissibility on S-algebraic groups satisfy the following. For every $g \in G$, there exists a positive constant $c(g) > 0$ such that $g\mathcal{O}_\varepsilon g^{-1}$ contains $\mathcal{O}_{c(g)\varepsilon}$ for all $0 < \varepsilon < \varepsilon_0$. We can therefore easily conclude the following.

Lemma 5.24. Stability under translations. *Let G_t be a one-parameter family of coarsely admissible averages on an S-algebraic group as in Definition 3.4. Then for any $g, h \in G$, the family gG_th is also coarsely admissible. If G_t is admissible, then so is gG_th.*

Proof. To see that for coarsely admissible averages G_t, the averages gG_th are also coarsely admissible, note that for any bounded set B,

$$BgG_thB \subset B'G_tB' \subset G_{t+c'}$$

and in addition $g^{-1}G_{t+c'}h^{-1} \subset G_{t+c''}$, so that

$$BgG_thB \subset G_{t+c'} \subset gG_{t+c''}h.$$

As to the second condition of coarse admissibility, by unimodularity,

$$m_G(gG_{t+c}h) \leq D \, m_G(gG_th)$$

for some $D > 0$, and so gG_th is coarsely admissible.

Now let G_t be an admissible one-parameter family and the elements $h, g \in G$ be fixed. For every open set $\mathcal{O}_\varepsilon$ in the basis, the set $g\mathcal{O}_\varepsilon g^{-1} \cap h^{-1}\mathcal{O}_\varepsilon h$ is open and contains $\mathcal{O}_{\eta(\varepsilon)}$. It follows from the definition of the basis $\mathcal{O}_\varepsilon$ (see Remark 3.6(2)) that we can choose $\eta(\varepsilon) \geq c_0\varepsilon$, for some fixed positive $c_0 = c_0(g, h) < 1$, uniformly for all $0 < \varepsilon < \varepsilon(g, h)$. Then, checking the conditions in the definition of admissibility,

$$\mathcal{O}_{\eta(\varepsilon)}gG_th\mathcal{O}_{\eta(\varepsilon)} \subset g\mathcal{O}_\varepsilon G_t\mathcal{O}_\varepsilon h \subset gG_{t+c\varepsilon}h,$$

so that for all $0 < \varepsilon < \varepsilon'(g, h)$,

$$\mathcal{O}_\varepsilon gG_th\mathcal{O}_\varepsilon \subset gG_{t+\varepsilon c/c_0}h.$$

When G is totally disconnected and G_t satisfies $KG_tK = G_t$, clearly gG_th is also invariant under translation by the compact open subgroup $K' = gKg^{-1} \cap hKh^{-1}$.

As to the Lipschitz continuity of the measure of the family, we have of course, since G is unimodular,

$$m_G(gG_{t+\varepsilon}h) = m_G(G_{t+\varepsilon}) \leq (1 + c\varepsilon)m_G(G_t) = (1 + c\varepsilon)m_G(gG_th).$$

This completes the proof of the lemma. $\square$

Chapter Six

Proof of ergodic theorems for lattice subgroups

We now turn to consideration of a discrete lattice subgroup Γ of an lcsc group G and to the problem of establishing ergodic theorems for actions of the lattice, given the validity of ergodic theorems for actions of G. We will begin by discussing the induced action and then develop a series of reduction arguments of increasing precision to achieve this goal. In §6.8 we put all the arguments together and finalize the proofs of all the ergodic theorems stated for G and Γ, and in §6.9 we prove the equidistribution result.

The existence of a lattice implies that G is unimodular, and we denote Haar measure on G by m_G as before. We denote by $m_{G/\Gamma}$ the corresponding measure on G/Γ. In the present chapter we normalize m_G so that $m_{G/\Gamma}(G/\Gamma) = 1$.

6.1 INDUCED ACTION

For a family of Borel subsets $\{G_t\}_{t>0}$, we consider the averages λ_t uniformly distributed on $\Gamma_t = G_t \cap \Gamma$. We will use the mean, maximal, and pointwise ergodic theorems established for the averages β_t acting in a G-action in order to establish similar ergodic theorems for the averages λ_t acting in a Γ-action. The fundamental link used to implement this reduction is the well-known construction of the G-action induced from a measure-preserving action of Γ, to which we now turn (see [Z] for a general discussion).

Thus let Γ act on a standard Borel space $(X, \mathcal{B}, \mu)$, preserving the probability measure μ. Let

$$\bar{Y} \overset{\text{def}}{=} G \times X.$$

Define the right action of Γ on $\bar{Y}$,

$$(g, x) \cdot \gamma = (g\gamma, \gamma^{-1}x), \quad (g, x) \in \bar{Y}, \ \gamma \in \Gamma, \tag{6.1}$$

and the left action of G,

$$g_1 \cdot (g, x) = (g_1 g, x), \quad (g, x) \in \bar{Y}, \ g_1 \in G. \tag{6.2}$$

The space $\bar{Y}$ is equipped with the product measure $m_G \otimes \mu$, which is preserved by these actions. Since the actions (6.1) and (6.2) commute, there is a well-defined action of G on the factor space

$$Y \overset{\text{def}}{=} \bar{Y}/\Gamma.$$

We denote by ι the projection map $\iota : \bar{Y} \to Y$. Note that Y admits a natural map $j : (g, x)\Gamma \mapsto g\Gamma$ onto G/Γ. This map is Borel-measurable and G-equivariant,

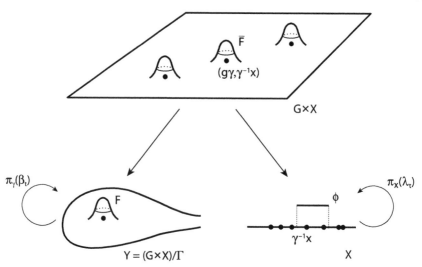

Figure 6.1 Ergodic theorem for Γ-actions

and thus Y is a bundle over the homogeneous space G/Γ, with the fiber over each point $g\Gamma$ identified with X.

For a bounded measurable function $\chi : G \to \mathbb{R}$ with compact support and a measurable function $\phi : X \to \mathbb{R}$, we define $\bar{F} : \bar{Y} \to \mathbb{R}$ by $\bar{F}(g, x) = \chi(g)\phi(x)$. We then define $F : Y \to \mathbb{R}$ by summing over Γ-orbits:

$$F(y) = F((g, x)\Gamma) = \sum_{\gamma \in \Gamma} \chi(g\gamma)\phi(\gamma^{-1}x) = \sum_{\gamma \in \Gamma} \bar{F}((g, x)\gamma). \qquad (6.3)$$

There is a unique G-invariant Borel measure ν on Y such that

$$\int_Y F \, d\nu = \left(\int_G \chi \, dm_G \right) \left(\int_X \phi \, d\mu \right). \qquad (6.4)$$

For F defined above, we have the following expression for the averaging operators β_t that we will consider below. Let $(h, x)\Gamma = y \in Y$, and then

$$\pi_Y(\beta_t)F(y) = \frac{1}{m_G(G_t)} \int_G F(g^{-1}y)d\beta_t(g)$$

$$= \sum_{\gamma \in \Gamma} \frac{1}{m_G(G_t)} \left(\int_{G_t} \chi(g^{-1}h\gamma)dm_G(g) \right) \phi(\gamma^{-1}x).$$

As noted in Chapter 6, the latter expression will serve as the basic link between the averaging operators β_t on G acting on $L^p(Y)$ and the averaging operators λ_t acting on $L^p(X)$.

We now recall the following fact regarding induced actions, which will play an important role below. Namely, it will allow us to deduce results about the pointwise behavior of the averages λ_t on the Γ-orbits in X from the pointwise behavior of the averages β_t on G-orbits in Y.

Consider the factor map $j : (Y, \nu) \to (G/\Gamma, m_{G/\Gamma})$, which is Borel-measurable, everywhere defined, G-equivariant, and measure-preserving. For a Lebesgue-measurable set $B \subset Y$, the set $B_{z\Gamma} = j^{-1}(z\Gamma) \cap B$ is a Lebesgue-measurable subset of X for almost every $y\Gamma \in G/\Gamma$ (see, e.g., [R, Thm. 7.12]). Recall that the Lebesgue σ-algebra is defined as the completion of the Borel σ-algebra w.r.t. the measure at hand, namely, ν on Y or μ on X.

Any set B can be written as the disjoint union $B = \bigsqcup_{z\Gamma \in G/\Gamma} B_{z\Gamma}$. Furthermore, the G-action is given by

$$(gB)_{gz\Gamma} = \alpha(g, z\Gamma) B_{z\Gamma},$$

where $\alpha : G \times G/\Gamma \to \Gamma$ is a Borel cocycle associated with a Borel section of the canonical projection $G \to G/\Gamma$. Clearly, if $B_{z\Gamma}$ is Lebesgue-measurable, so is $\alpha(g, z\Gamma) B_{z\Gamma} = (gB)_{gz\Gamma}$.

We can now state the following fact, whose proof is included for completeness.

Lemma 6.1. *If $B \subset Y$ is a Lebesgue-measurable set with $\nu(B) = 1$, which is strictly G-invariant ($gB = B$ for all $g \in G$), then $B_{z\Gamma} \subset X$ is Lebesgue-measurable and $\mu(B_{z\Gamma}) = 1$, for **every** $z\Gamma \in G/\Gamma$ (and not only for almost every $z\Gamma$).*

Proof. $B_{z\Gamma} \subset X$ is Lebesgue-measurable for almost every $z\Gamma$, as noted above. But since $gB = B$, $B_{gz\Gamma}$ is the image of $B_{z\Gamma}$ under the measurable automorphism $\alpha(g, z\Gamma)$, so it also Lebesgue-measurable. Since G/Γ is a transitive G-space, $B_{z\Gamma}$ is Lebesgue-measurable for **every** $z\Gamma \in G/\Gamma$. The map $b : G/\Gamma \to \mathbb{R}_+$ given by $z\Gamma \mapsto \mu(B_{z\Gamma})$ is everywhere-defined, Lebesgue-measurable, and strictly G-invariant, namely, $b(gz\Gamma) = b(z\Gamma)$ for all $g \in G$ and $z\Gamma \in G/\Gamma$ since Γ preserves the measure μ. Again, since G/Γ is a transitive G-space, $b(z\Gamma)$ is strictly a constant, and this constant is of course 1. $\qquad\square$

We conclude our discussion of induced actions with the following simple and easily verifiable facts.

Lemma 6.2. *Let $1 \le p \le \infty$ and Q be a compact subset of G.*

1. *There exists $a_{p,Q} > 0$ such that for every $\phi \in L^p(\mu)$ and a bounded $\chi : G \to \mathbb{R}$ such that $\mathrm{supp}(\chi) \subset Q$, with F defined as in (6.3),*

$$\|F\|_{L^p(\nu)} \le a_{p,Q} \cdot \|\chi\|_{L^p(m_G)} \cdot \|\phi\|_{L^p(\mu)}.$$

Moreover, if Q is contained in a sufficiently small neighborhood of e, then

$$\|F\|_{L^p(\nu)} = \|\chi\|_{L^p(m_G)} \cdot \|\phi\|_{L^p(\mu)}.$$

2. *There exists $b_{p,Q} > 0$ such that for any measurable $F : Y \to \mathbb{R}$,*

$$\|F \circ \iota\|_{L^p(m_G \otimes \mu|_{Q \times X})} \le b_{p,Q} \cdot \|F\|_{L^p(\nu)}.$$

Here $\iota : G \times X \to Y$ is the projection map. When $Q = \mathcal{O}_\varepsilon$, we denote $b_{p,Q} = b_{p,\varepsilon}$.

6.2 REDUCTION THEOREMS

We now turn to formulating the fundamental result reducing the ergodic theory of the lattice subgroup Γ to that of the enveloping group G.

Such a result necessarily involves an approximation argument based on smoothing, and thus the metric properties of a shrinking family of neighborhoods in G come into play. The crucial property is finiteness of the upper local dimension of G (see Definition 3.5), namely,

$$\varrho_0 \stackrel{\text{def}}{=} \limsup_{\varepsilon \to 0^+} \frac{\log m_G(\mathcal{O}_\varepsilon)}{\log \varepsilon} < \infty.$$

We will assume this condition when considering admissible sets throughout our discussion below. Note that for S-algebraic groups as in Definition 3.4, and for the sets $\mathcal{O}_\varepsilon$ we chose in that case, ϱ_0 is simply the real dimension of the Archimedean factor and thus vanishes for totally disconnected groups.

Let us note that the induced representation of G on $L^p(Y)$, $1 \le p \le \infty$, contains the representation of G on $L^p(G/\Gamma)$ as a subrepresentation. Thus, whenever a strong maximal inequality, exponential strong maximal inequality, norm decay estimate, spectral gap condition, or mean or pointwise ergodic theorem holds for $\pi_Y(\beta_t)$ acting on $L^p(Y)$, it also holds for $\pi_{G/\Gamma}(\beta_t)$ acting on $L^p(G/\Gamma)$.

We now formulate the following reduction theorem and emphasize that it is valid *for every discrete lattice subgroup of every lcsc group.*

Recall that the Haar measure m_G on G is normalized so that the induced measure of G/Γ is 1, and that (Y, v) is the G-action induced from the Γ-action on (X, μ).

Theorem 6.3. Reduction theorem. *Let G be an lcsc group, $\mathcal{O}_\varepsilon$ of finite upper local dimension, G_t an increasing family of bounded Borel sets, and Γ a lattice subgroup. Let $p \ge r \ge 1$ and consider the averages β_t on G_t and λ_t on $\Gamma \cap G_t$ as above. Then,*

1. *If the family $\{G_t\}_{t>0}$ is coarsely admissible, the strong maximal inequality for β_t in $(L^p(v), L^r(v))$ implies the strong maximal inequality for λ_t in $(L^p(\mu), L^r(\mu))$.*

2. *If the family $\{G_t\}_{t>0}$ is admissible, the mean ergodic theorem for β_t in $L^p(v)$ implies the mean ergodic theorem for λ_t in $L^p(\mu)$.*

3. *If the family $\{G_t\}_{t>0}$ is quasi-uniform and the pointwise ergodic theorem holds for β_t in $L^p(v)$, the pointwise ergodic theorem holds for λ_t in $L^p(\mu)$.*

4. *If the family $\{G_t\}_{t>0}$ is admissible and $r > \varrho_0$, the exponential mean ergodic theorem for β_t in $(L^p(v), L^r(v))$ implies the exponential mean ergodic theorem for λ_t in $(L^p(\mu), L^r(\mu))$ (but the rate may change).*

5. *Let the family $\{G_t\}_{t>0}$ be admissible, let $p \ge r > \varrho_0$, and assume β_t satisfies the exponential mean ergodic theorem in $(L^p(v), L^r(v))$, as well as the strong maximal inequality in $L^q(v)$, for $q > 1$. Then λ_t satisfies the exponential strong maximal inequality in (L^v, L^w) with v, w such that $1/v = (1 - u)/q$ and $1/w = (1 - u)/q + u/r$ for some $u \in (0, 1)$.*

6. *Under the assumptions of part 5, λ_t satisfies the exponential pointwise ergodic theorem in (L^v, L^w) with v, w such that $1/v = (1 - u)/q$ and $1/w = (1 - u)/q + u/r$ for some $u \in (0, 1)$.*

One basic ingredient in the proof of Theorem 6.3 is as follows.

Theorem 6.4. Lattice points. *Let G, G_t, β_t, and λ_t be as in Theorem 6.3. Then,*

1. *Suppose that the family $\{G_t\}_{t>0}$ is coarsely admissible and β_t satisfies the strong maximal inequality in $(L^p(m_{G/\Gamma}), L^r(m_{G/\Gamma}))$ for some $p \geq r \geq 1$. Then for some $C > 0$ and all sufficiently large t,*

$$C^{-1} \cdot m_G(G_t) \leq |\Gamma \cap G_t| \leq C \cdot m_G(G_t).$$

2. *Suppose that the family $\{G_t\}_{t>0}$ is admissible and β_t satisfies the mean ergodic theorem in $L^p(m_{G/\Gamma})$ for some $p \geq 1$. Then*

$$\lim_{t \to \infty} \frac{|\Gamma \cap G_t|}{m_G(G_t)} = 1.$$

3. *Suppose that the family $\{G_t\}_{t>0}$ is quasi-uniform and β_t satisfies the pointwise ergodic theorem in $L^\infty(m_{G/\Gamma})$. Then*

$$\lim_{t \to \infty} \frac{|\Gamma \cap G_t|}{m_G(G_t)} = 1.$$

4. *Suppose that the family $\{G_t\}_{t>0}$ is admissible and β_t satisfies the exponential mean ergodic theorem in $(L^p(m_{G/\Gamma}), L^r(m_{G/\Gamma}))$ for some $p \geq r \geq 1$. Then for some $\alpha > 0$ (made explicit in §8.1 below),*

$$\frac{|\Gamma \cap G_t|}{m_G(G_t)} = 1 + O(e^{-\alpha t}).$$

The proofs of Theorems 6.3 and 6.4 will be divided into a sequence of separate arguments, which will occupy §6.3–§6.7. We will freely use the notation from §6.1.

6.3 STRONG MAXIMAL INEQUALITY

In this subsection we assume that the family $\{G_t\}_{t>0}$ is coarsely admissible and as usual set $\Gamma_t = G_t \cap \Gamma$. We start with the following basic estimate.

Lemma 6.5.

1. $|\Gamma_t| \leq C m_G(G_t)$.

2. *Assuming the strong maximal inequality in $(L^p(m_{G/\Gamma}), L^r(m_{G/\Gamma}))$ for the averages β_t for some $p \geq r \geq 1$, we have $|\Gamma_t| \geq C' m_G(G_t)$ for sufficiently large t.*

Proof. Let $B \subset G$ be a bounded measurable subset of positive measure and assume that B is small enough so that all of its right translates by elements of Γ are pairwise disjoint. Then by (3.3) and (3.4),

$$|\Gamma_t| = \frac{1}{m_G(B)} \sum_{\gamma \in \Gamma_t} m_G(B\gamma) = \frac{1}{m_G(B)} m_G \left(\bigcup_{\gamma \in \Gamma_t} B\gamma \right)$$

$$\leq \frac{1}{m_G(B)} m_G(G_{t+c}) \leq C m_G(G_t).$$

This proves part 1 of the lemma.

To prove part 2, we first show the following.

Claim. *There exist a compact set $Q \subset G/\Gamma$ and $x_0 \in G/\Gamma$ such that*

$$\liminf_{t \to \infty} \pi_{G/\Gamma}(\beta_t) \chi_Q(x_0) = \liminf_{t \to \infty} \frac{m_G(\{g \in G_t : gx_0 \in Q\})}{m_G(G_t)} > 0.$$

Proof. Suppose that the claim is false. For a compact set $Q \subset G/\Gamma$, denote by ψ the characteristic function of the set $(G/\Gamma) \setminus Q$, the complement of Q. Then for every $x \in G/\Gamma$,

$$\sup_{t \geq t_0} \pi_{G/\Gamma}(\beta_t) \psi(x) \geq \limsup_{t \to \infty} \pi_{G/\Gamma}(\beta_t) \psi(x) = 1.$$

On the other hand,

$$\|\psi\|_{L^p(G/\Gamma)} = m_{G/\Gamma}((G/\Gamma) \setminus Q)^{1/p},$$

and it can be made arbitrary small by increasing Q. This contradicts the strong maximal inequality and proves the claim.   □

Continuing with the proof of Lemma 6.5, denote by χ_Q the characteristic function of the set Q. Then for some $x_0 \in G/\Gamma$,

$$\liminf_{t \to \infty} \pi_{G/\Gamma}(\beta_t) \chi_Q(x_0) = C_0 > 0.$$

There exists a nonnegative measurable function $\bar{\chi} : G \to \mathbb{R}$ with compact support such that

$$\chi_Q(g\Gamma) = \sum_{\gamma \in \Gamma} \bar{\chi}(g\gamma)$$

since the projection $C_c^+(G) \to C_c^+(G/\Gamma)$ by summing over Γ-orbits is onto.

Letting $x_0 = g_0\Gamma$, we conclude that

$$\int_{G_t} \sum_{\gamma \in \Gamma} \bar{\chi}(g^{-1}g_0\gamma) dm_G(g) \geq \frac{1}{2} C_0 m_G(G_t)$$

for sufficiently large t. Now, if $\bar{\chi}(g^{-1}g_0\gamma) \neq 0$ for some $g \in G_t$, then

$$\gamma \in g_0^{-1} \cdot G_t \cdot (\operatorname{supp} \bar{\chi}) \subset G_{t+c}$$

by (3.3). Hence,

$$\int_{G_t} \sum_{\gamma \in \Gamma} \bar{\chi}(g^{-1}g_0\gamma) \, dm_G(g) \leq \sum_{\gamma \in \Gamma_{t+c}} \int_{G_t^{-1}g_0\gamma} \bar{\chi} \, dm_G$$

$$\leq |\Gamma_{t+c}| \cdot \int_G \bar{\chi} \, dm_G.$$

Now Lemma 6.5 follows from (3.4).   □

We now prove the following result, which reduces the strong maximal inequality for λ_t to the strong maximal inequality for β_t, under the assumption of coarse admissibility.

Theorem 6.6. *Suppose that β_t satisfies the strong maximal inequality in $(L^p(\nu), L^r(\nu))$, then the averages λ_t satisfy the strong maximal inequality in $(L^p(\mu), L^r(\mu))$.*

Proof. Take $\phi \in L^p(\mu)$, assumed real without loss of generality. Observe that it suffices to prove the theorem for $\phi \geq 0$. Write

$$\phi = \phi^+ - \phi^-,$$

where $\phi^+, \phi^- : X \to \mathbb{R}_+$ are Borel functions such that

$$\max\{\phi^+, \phi^-\} \leq |\phi|.$$

Assuming that the strong maximal inequality holds for both $\pi_X(\lambda_t)\phi^+$ and $\pi_X(\lambda_t)\phi^-$, we have

$$\left\| \sup_{t \geq t_0} |\pi_X(\lambda_t)\phi| \right\|_{L^r(\mu)} \leq \left\| \sup_{t \geq t_0} |\pi_X(\lambda_t)\phi^+| \right\|_{L^r(\mu)} + \left\| \sup_{t \geq t_0} |\pi_X(\lambda_t)\phi^-| \right\|_{L^r(\mu)}$$

$$\leq C\|\phi^+\|_{L^p(\mu)} + C\|\phi^-\|_{L^p(\mu)} \leq 2C\|\phi\|_{L^p(\mu)}.$$

Hence, we can assume that $\phi \geq 0$.

Let B be a positive-measure compact subset of G small enough so that all of its right translates under Γ are disjoint, let

$$\chi = \frac{\chi_B}{m_G(B)},$$

and let $F : Y \to \mathbb{R}$ be defined as in (6.3).

Claim. *There exist $c, D > 0$ such that for all sufficiently large t and every $h \in B$ and $x \in X$,*

$$\pi_X(\lambda_t)\phi(x) \leq D \cdot \pi_Y(\beta_{t+c})F(\iota(h, x)).$$

Proof. For $(h, x) \in G \times X$, we have (recall $\iota(h, x) = (h, x)\Gamma \in Y$)

$$\pi_Y(\beta_{t+c})F(\iota(h, x)) = \frac{1}{m_G(G_{t+c})} \int_{G_{t+c}} \left(\sum_{\gamma \in \Gamma} \chi(g^{-1}h\gamma)\phi(\gamma^{-1} \cdot x) \right) dm_G(g)$$

$$= \frac{1}{m_G(G_{t+c})} \sum_{\gamma \in \Gamma} \left(\int_{G_{t+c}} \chi(g^{-1}h\gamma) dm_G(g) \right) \phi(\gamma^{-1} \cdot x).$$

By (3.3), for $\gamma \in \Gamma_t$ and $h \in B$,

$$\mathrm{supp}(g \mapsto \chi(g^{-1}h\gamma)) = h\gamma \, \mathrm{supp}(\chi)^{-1} \subset G_{t+c}.$$

Hence,

$$\int_{G_{t+c}} \chi(g^{-1}h\gamma) dm_G(g) = 1.$$

Also, by Lemma 6.5 and (3.4),

$$|\Gamma_t| \geq C' m_G(G_{t+c}).$$

Thus summing only on $\gamma \in \Gamma_t$, so we can conclude that for $(h, x) \in B \times X$,

$$\pi_Y(\beta_{t+c})F(\iota(h, x)) \geq \frac{1}{m_G(G_{t+c})} \sum_{\gamma \in \Gamma_t} \left(\int_{G_{t+c}} \chi(g^{-1}h\gamma) \, dm_G(g) \right) \phi(\gamma^{-1} \cdot x)$$

$$= \frac{1}{m_G(G_{t+c})} \sum_{\gamma \in \Gamma_t} \phi(\gamma^{-1} \cdot x) \geq C'' \pi_X(\lambda_t)\phi(x).$$

This proves the claim.   □

Continuing with the proof of Theorem 6.6, we would now like to take the supremum over t on both sides of the inequality just proved in the claim. Let us lift $\pi_X(\lambda_t)\phi$ and view it as a function defined on $B \times X$ depending only the second coordinate. We can view $\pi_Y(\beta_{t+c})(F \circ \iota)$ as defined on $B \times X$ as well, and by the claim, for sufficiently large $t_0' > 0$, we obtain by integration

$$\left\| \sup_{t \geq t_0'} |\pi_X(\lambda_t)\phi)| \right\|_{L^r(\mu)} = m_G(B)^{-1/r} \left\| \sup_{t \geq t_0'} |\pi_X(\lambda_t)\phi| \right\|_{L^r(m_G \otimes \mu|_{B \times X})}$$

$$\leq C' \left\| \sup_{t \geq t_0'} |\pi_Y(\beta_{t+c})(F \circ \iota)| \right\|_{L^r(m_G \otimes \mu|_{B \times X})}.$$

Hence, by Lemma 6.2(2) and the strong $(L^p(\nu), L^r(\nu))$ maximal inequality for β_t,

$$\left\| \sup_{t \geq t_0'} |\pi_X(\lambda_t)\phi)| \right\|_{L^r(\mu)} \leq C' b_{r, B} \left\| \sup_{t \geq t_0'} |\pi_Y(\beta_{t+c})F| \right\|_{L^r(\nu)} \leq C'' \|F\|_{L^p(\nu)}$$

$$= C'' \|\chi\|_{L^p(m_G)} \cdot \|\phi\|_{L^p(\mu)} \leq C \|\phi\|_{L^p(\mu)},$$

where the equality uses the fact that B has disjoint right translates under Γ and Lemma 6.2(1).

This concludes the proof of Theorem 6.6.   □

6.4 MEAN ERGODIC THEOREM

We now turn from maximal inequalities to establishing convergence results for averages on Γ, using smoothing to approximate discrete averages by absolutely continuous ones and thus utilizing the finiteness of the upper local dimension of G. To elucidate the exact condition that will come into play, let us generalize the definition of upper local dimension somewhat and consider a base of neighborhoods $\{\mathcal{O}_\varepsilon\}_{0 < \varepsilon < 1}$ of e in G such that $\mathcal{O}_\varepsilon$ are symmetric, bounded, and decreasing with ε. We assume that the family $\{G_t\}_{t>0}$ satisfies the following conditions:

- There exists $c > 0$ such that for every small $\varepsilon > 0$ and $t \geq t(\varepsilon)$,

$$\mathcal{O}_\varepsilon \cdot G_t \cdot \mathcal{O}_\varepsilon \subset G_{t+c\varepsilon}. \tag{6.5}$$

- Setting

$$\delta_\varepsilon = \limsup_{t \to \infty} \frac{m_G(G_{t+\varepsilon} - G_t)}{m_G(G_t)},$$

we have, for some $p \geq 1$,

$$\delta_\varepsilon^p \cdot m_G(\mathcal{O}_\varepsilon)^{-1} \to 0 \quad \text{as } \varepsilon \to 0^+. \tag{6.6}$$

Note that if the family $\{G_t\}_{t>0}$ is admissible and

$$\varrho_0 = \limsup_{\varepsilon \to 0^+} \frac{\log m_G(\mathcal{O}_\varepsilon)}{\log \varepsilon} < \infty,$$

then (6.6) holds for $p > \varrho_0$.

Note that (6.6) implies that $\delta_\varepsilon \to 0$ as $\varepsilon \to 0^+$. For every $\delta > \delta_\varepsilon$ and for sufficiently large t,

$$m_G(G_{t+\varepsilon}) \leq (1+\delta) m_G(G_t). \tag{6.7}$$

We start with the following result identifying the main term in the lattice point-counting problem, which is a necessary prerequisite for the mean ergodic theorem.

Lemma 6.7. *Under conditions (6.5) and (6.6), if the mean ergodic theorem holds for β_t in $L^q(m_{G/\Gamma})$ for some $q \geq 1$, then*

$$|\Gamma_t| \sim m_G(G_t) \quad \text{as } t \to \infty.$$

Proof. Let

$$\chi_\varepsilon = \frac{\chi_{\mathcal{O}_\varepsilon}}{m_G(\mathcal{O}_\varepsilon)}$$

and

$$\phi_\varepsilon(g\Gamma) = \sum_{\gamma \in \Gamma} \chi_\varepsilon(g\gamma).$$

Note that ϕ_ε is a measurable bounded function on G/Γ with compact support,

$$\int_G \chi_\varepsilon \, dm_G = 1, \quad \text{and} \quad \int_{G/\Gamma} \phi_\varepsilon \, dm_{G/\Gamma} = 1.$$

It follows from the mean ergodic theorem that for every $\delta > 0$,

$$m_{G/\Gamma}(\{g\Gamma \in G/\Gamma : |\pi_{G/\Gamma}(\beta_t)\phi_\varepsilon(g\Gamma) - 1| > \delta\}) \to 0 \quad \text{as } t \to \infty.$$

It follows, in particular, that for sufficiently large t there exist $g_t \in \mathcal{O}_\varepsilon$ such that $|\pi_{G/\Gamma}(\beta_t)\phi_\varepsilon(g_t\Gamma) - 1| \leq \delta$, or equivalently,

$$1 - \delta \leq \frac{1}{m_G(G_t)} \int_{G_t} \phi_\varepsilon(g^{-1}g_t\Gamma) dm_G(g) \leq 1 + \delta. \tag{6.8}$$

Thus let us now prove the following.

Claim. *For small $\varepsilon > 0$, sufficiently large t, and every $h \in \mathcal{O}_\varepsilon$,*

$$\int_{G_{t-c\varepsilon}} \phi_\varepsilon(g^{-1}h\Gamma) \, dm_G(g) \leq |\Gamma_t| \leq \int_{G_{t+c\varepsilon}} \phi_\varepsilon(g^{-1}h\Gamma) \, dm_G(g).$$

Proof. Indeed, if $\chi_\varepsilon(g^{-1}h\gamma) \neq 0$ for some $g \in G_{t-c\varepsilon}$ and $h \in \mathcal{O}_\varepsilon$, then by (6.5),

$$\gamma \in h^{-1} \cdot G_{t-c\varepsilon} \cdot (\text{supp } \chi_\varepsilon) \subset G_t.$$

Hence,

$$\int_{G_{t-c\varepsilon}} \phi_\varepsilon(g^{-1}h\Gamma)\, dm_G(g) \leq \sum_{\gamma \in \Gamma_t} \int_{G_t} \chi_\varepsilon(g^{-1}h\gamma)\, dm_G(g) \leq |\Gamma_t|.$$

In the other direction, for $\gamma \in \Gamma_t$ and $h \in \mathcal{O}_\varepsilon$,

$$\text{supp}(g \mapsto \chi_\varepsilon(g^{-1}h\gamma)) = h\gamma(\text{supp } \chi_\varepsilon)^{-1} \subset G_{t+c\varepsilon}.$$

Therefore, since $\chi_\varepsilon \geq 0$,

$$\int_{G_{t+c\varepsilon}} \phi_\varepsilon(g^{-1}h\Gamma)\, dm_G(g) \geq \sum_{\gamma \in \Gamma_t} \int_{G_{t+c\varepsilon}} \chi_\varepsilon(g^{-1}h\gamma)\, dm_G(g) \geq |\Gamma_t|,$$

and this establishes the claim.   □

Continuing with the proof of Lemma 6.7, let us take $h = g_t$ defined above. By the claim and (6.8),

$$|\Gamma_t| \leq (1+\delta)m_G(G_{t+\varepsilon}),$$

and the upper estimate on $|\Gamma_t|$ follows from (6.7). The lower estimate is proved similarly.   □

We now generalize Lemma 6.7 and prove the following result, reducing the mean ergodic theorem for λ_t in an arbitrary action to the mean ergodic theorem for β_t in the induced action.

Theorem 6.8. *Under conditions (6.5) and (6.6), if the mean ergodic theorem holds for β_t in $L^p(\nu)$, then the mean ergodic theorem holds for λ_t in $L^p(\mu)$.*

Proof. Take small $\varepsilon > 0$ and $\delta \in (\delta_\varepsilon, 1)$, where δ_ε (as well as p) are defined by (6.6).

We need to show that for every $\phi \in L^p(\mu)$,

$$\left\| \pi_X(\lambda_t)\phi - \int_X \phi\, d\mu \right\|_{L^p(\mu)} \to 0 \quad \text{as } t \to \infty,$$

and without loss of generality, we may assume that $\phi \geq 0$. Let

$$\chi_\varepsilon = \frac{\chi_{\mathcal{O}_\varepsilon}}{m_G(\mathcal{O}_\varepsilon)}$$

and let $F_\varepsilon : Y \to \mathbb{R}$ be defined using χ_ε and ϕ as in (6.3). Then $F_\varepsilon \in L^p(\nu)$ and

$$\int_Y F_\varepsilon\, d\nu = \int_X \phi\, d\mu.$$

Step 1. *For every $(h, x) \in \mathcal{O}_\varepsilon \times X$ and sufficiently large t,*

$$(1+\delta)^{-1}\pi_Y(\beta_{t-c\varepsilon})F_\varepsilon(\iota(h, x)) \leq \pi_X(\lambda_t)\phi(x) \leq (1+\delta)\pi_Y(\beta_{t+c\varepsilon})F_\varepsilon(\iota(h, x)).$$

To prove the first inequality, note that by Lemma 6.7 and (6.7),

$$(1+\delta)^{-1} m_G(G_{t+c\varepsilon}) \le |\Gamma_t| \le (1+\delta) m_G(G_{t-c\varepsilon})$$

for suffciently large t.

For $(h, x) \in G \times X$,

$$\pi_Y(\beta_t) F_\varepsilon(\iota(h, x)) = \frac{1}{m_G(G_t)} \int_{G_t} \left(\sum_{\gamma \in \Gamma} \chi_\varepsilon(g^{-1} h\gamma) \phi(\gamma^{-1} \cdot x) \right) dm_G(g)$$

$$= \frac{1}{m_G(G_t)} \sum_{\gamma \in \Gamma} \left(\int_{G_t} \chi_\varepsilon(g^{-1} h\gamma) \, dm_G(g) \right) \phi(\gamma^{-1} \cdot x).$$

If $\chi_\varepsilon(g^{-1} h\gamma) \ne 0$ for some $g \in G_t$ and $h \in \mathcal{O}_\varepsilon$, then by (6.5),

$$\gamma \in h^{-1} g \, \mathrm{supp}(\chi_\varepsilon) \subset G_{t+c\varepsilon}.$$

Using

$$\int_G \chi_\varepsilon \, dm_G = 1 \quad \text{and} \quad \chi_\varepsilon \ge 0, \tag{6.9}$$

we deduce that for $(h, x) \in \mathcal{O}_\varepsilon \times X$,

$$\pi_Y(\beta_t) F_\varepsilon(\iota(h, x)) = \frac{1}{m_G(G_t)} \sum_{\gamma \in \Gamma_{t+c\varepsilon}} \left(\int_{G_t} \chi_\varepsilon(g^{-1} h\gamma) \, dm_G(g) \right) \phi(\gamma^{-1} \cdot x)$$

$$\le \frac{1}{m_G(G_t)} \sum_{\gamma \in \Gamma_{t+c\varepsilon}} \phi(\gamma^{-1} \cdot x) \le (1+\delta) \pi_X(\lambda_{t+c\varepsilon}) \phi(x).$$

To prove the second inequality, note that by (6.5), for $\gamma \in \Gamma_{t-c\varepsilon}$ and $h \in \mathcal{O}_\varepsilon$,

$$\mathrm{supp}(g \mapsto \chi_\varepsilon(g^{-1} h\gamma)) = h\gamma \, \mathrm{supp}(\chi_\varepsilon)^{-1} \subset G_t.$$

By (6.9), this implies that for $(h, x) \in \mathcal{O}_\varepsilon \times X$,

$$\pi_Y(\beta_t) F_\varepsilon(\iota(h, x)) \ge \frac{1}{m_G(G_t)} \sum_{\gamma \in \Gamma_{t-c\varepsilon}} \left(\int_{G_t} \chi_\varepsilon(g^{-1} h\gamma) \, dm_G(g) \right) \phi(\gamma^{-1} \cdot x)$$

$$= \frac{1}{m_G(G_t)} \sum_{\gamma \in \Gamma_{t-c\varepsilon}} \phi(\gamma^{-1} \cdot x) \ge (1+\delta)^{-1} \pi_X(\lambda_{t-c\varepsilon}) \phi(x).$$

Using Lemma 6.7 and (6.7) again and then shifting the indices of β_t completes the proof of step 1.

We now continue with the proof of Theorem 6.8. To simplify the notation, we write for a measurable function $\Psi : Y \to \mathbb{R}$,

$$\|\Psi\|_{p,\varepsilon} \overset{\text{def}}{=} \|\Psi \circ \iota\|_{L^p(m_G \otimes \mu|_{\mathcal{O}_\varepsilon \times X})}.$$

Now by Lemma 6.2(2), for each fixed $\varepsilon > 0$,

$$\|\Psi\|_{p,\varepsilon} \le b_{p,\varepsilon} \|\Psi\|_{L^p(\nu)}, \tag{6.10}$$

and clearly, if $\varepsilon' < \varepsilon$, we may take $b_{p,\varepsilon'} \le b_{p,\varepsilon}$.

Step 2. *For every sufficiently small fixed $\varepsilon > 0$,*

$$\|\pi_Y(\beta_{t+c\varepsilon})F_\varepsilon - \pi_Y(\beta_t)F_\varepsilon\|_{p,\varepsilon} \to 0 \quad as \ t \to \infty$$

and

$$\limsup_{t\to\infty} \|\pi_Y(\beta_t)F_\varepsilon\|_{p,\varepsilon} \leq b_{p,\varepsilon}\|\phi\|_{L^1(\mu)}.$$

For the proof, let us note that by the triangle inequality and (6.10),

$$\|\pi_Y(\beta_{t+c\varepsilon})F_\varepsilon - \pi_Y(\beta_t)F_\varepsilon\|_{p,\varepsilon}$$

$$\leq b_{p,\varepsilon}\left\|\pi_Y(\beta_{t+c\varepsilon})F_\varepsilon - \int_Y F_\varepsilon \, dv\right\|_{L^p(v)} + b_{p,\varepsilon}\left\|\pi_Y(\beta_t)F_\varepsilon - \int_Y F_\varepsilon \, dv\right\|_{L^p(v)}.$$

Similarly,

$$\|\pi_Y(\beta_t)F_\varepsilon\|_{p,\varepsilon} \leq b_{p,\varepsilon}\left\|\pi_Y(\beta_t)F_\varepsilon - \int_Y F_\varepsilon \, dv\right\|_{L^p(v)} + b_{p,\varepsilon}\int_Y F_\varepsilon \, dv$$

$$= b_{p,\varepsilon}\left\|\pi_Y(\beta_t)F_\varepsilon - \int_Y F_\varepsilon \, dv\right\|_{L^p(v)} + b_{p,\varepsilon}\int_X \phi \, d\mu.$$

Hence, step 2 follows from the mean ergodic theorem for β_t in $L^p(v)$.

To complete the proof of Theorem 6.8, we need to estimate

$$\left\|\pi_X(\lambda_t)\phi - \int_X \phi \, d\mu\right\|_{L^p(\mu)}$$

$$= m_G(\mathcal{O}_\varepsilon)^{-1/p}\left\|\pi_X(\lambda_t)\phi - \int_X \phi \, d\mu\right\|_{L^p(m_G\otimes\mu|_{\mathcal{O}_\varepsilon\times X})},$$

where we have lifted $\pi_X(\lambda_t)\phi$ to a function on $\mathcal{O}_\varepsilon \times X$ depending only on the second variable. By the triangle inequality,

$$\left\|\pi_X(\lambda_t)\phi - \int_X \phi \, d\mu\right\|_{L^p(m_G\otimes\mu|_{\mathcal{O}_\varepsilon\times X})}$$

$$\leq \left\|\pi_X(\lambda_t)\phi - (1+\delta)^{-1}\pi_Y(\beta_{t-c\varepsilon})(F_\varepsilon \circ \iota)\right\|_{L^p(m_G\otimes\mu|_{\mathcal{O}_\varepsilon\times X})}$$

$$+ \left\|(1+\delta)^{-1}\pi_Y(\beta_{t-c\varepsilon})(F_\varepsilon \circ \iota) - \int_X \phi \, d\mu\right\|_{L^p(m_G\otimes\mu|_{\mathcal{O}_\varepsilon\times X})}.$$

We estimate the two summands as follows. First, using step 1, we estimate the first summand by

$$\left\|\pi_X(\lambda_t)\phi - (1+\delta)^{-1}\pi_Y(\beta_{t-c\varepsilon})(F_\varepsilon \circ \iota)\right\|_{L^p(m_G\otimes\mu|_{\mathcal{O}_\varepsilon\times X})}$$

$$\leq \left\|(1+\delta)\pi_Y(\beta_{t+c\varepsilon})F_\varepsilon - (1+\delta)^{-1}\pi_Y(\beta_{t-c\varepsilon})F_\varepsilon\right\|_{p,\varepsilon}$$

$$\leq \|\pi_Y(\beta_{t+c\varepsilon})F_\varepsilon - \pi_Y(\beta_{t-c\varepsilon})F_\varepsilon\|_{p,\varepsilon}$$

$$+ \delta\left(\|\pi_Y(\beta_{t+c\varepsilon})F_\varepsilon\|_{p,\varepsilon} + \|\pi_Y(\beta_{t-c\varepsilon})F_\varepsilon\|_{p,\varepsilon}\right).$$

Hence, it follows from step 2 and the mean ergodic theorem for β_t in $L^p(\nu)$ that the first summand is estimated by

$$\limsup_{t \to \infty} \left\| \pi_X(\lambda_t)\phi - (1+\delta)^{-1}\pi_Y(\beta_{t-c\varepsilon})(F_\varepsilon \circ \iota) \right\|_{L^p(m_G \otimes \mu|_{\mathcal{O}_\varepsilon \times X})}$$
$$\leq 2b_{p,\varepsilon}\delta\|\phi\|_{L^1(\mu)}.$$

As to the second summand, observing that for $\delta < 1$

$$\left\| (1+\delta)^{-1}\pi_Y(\beta_{t-c\varepsilon})F_\varepsilon - \int_X \phi\, d\mu \right\|_{p,\varepsilon}$$
$$\leq \left\| \pi_Y(\beta_{t-c\varepsilon})F_\varepsilon - \int_X \phi\, d\mu \right\|_{p,\varepsilon} + 2\delta \left\| \pi_Y(\beta_{t-c\varepsilon})F_\varepsilon \right\|_{p,\varepsilon},$$

we deduce from step 2 that the second summand is estimated by

$$\limsup_{t \to \infty} \left\| (1+\delta)^{-1}\pi_Y(\beta_{t-c\varepsilon})F_\varepsilon - \int_X \phi\, d\mu \right\|_{p,\varepsilon} \leq 4b_{p,\varepsilon}\delta\|\phi\|_{L^1(\mu)},$$

where $b_{p,\varepsilon}$ are uniformly bounded.

We have thus shown that for every $\delta \in (\delta_\varepsilon, 1)$ and a constant B independent of δ and ε,

$$\limsup_{t \to \infty} \left\| \pi_X(\lambda_t)\phi - \int_X \phi\, d\mu \right\|_{L^p(\mu)} \leq B\delta m_G(\mathcal{O}_\varepsilon)^{-1/p}\|\phi\|_{L^1(\mu)}.$$

By our choice of δ_ε in (6.6), this concludes the proof of Theorem 6.8. $\square$

6.5 POINTWISE ERGODIC THEOREM

In this section, we assume only that the family $\{G_t\}_{t>0}$ is quasi-uniform (see Definition 3.20). Recall that we showed in Corollary 5.23 that $gG_t h$ then satisfy the pointwise ergodic theorem if G_t does and that quasi-uniform families satisfy the invariance principle of Theorem 5.22, which plays a crucial role below.

Lemma 6.9. *Suppose that the pointwise ergodic theorem holds in $L^\infty(G/\Gamma)$ for the quasi-uniform family $\{gG_t\}$ for every $g \in G$. Then*

$$|\Gamma_t| \sim m_G(G_t) \quad as\ t \to \infty.$$

Proof. Let f be any measurable bounded function on G/Γ. G/Γ being a homogeneous G-space, it follows from Theorem 5.22 that the pointwise ergodic theorem holds for *every point* in G/Γ. In particular, this holds for the point $e\Gamma$, and so

$$\frac{1}{m_G(G_t)} \int_{G_t} f(g^{-1}\Gamma)dm_G(g) \to \int_{G/\Gamma} f\, dm_{G/\Gamma}$$

for every measurable bounded f. The lemma is then proved as Proposition 6.1 in [GW]. $\square$

We now come to the pointwise ergodic theorem for λ_t.

Theorem 6.10. *If the pointwise ergodic theorem holds for the quasi-uniform family β_t in $L^p(\nu)$, then the pointwise ergodic theorem holds for λ_t in $L^p(\mu)$.*

Proof. We need to show that for every $\phi \in L^p(\mu)$,

$$\pi_X(\lambda_t)\phi(x) \to \int_X \phi \, d\mu \quad \text{as } t \to \infty$$

for μ–almost everywhere $x \in X$, and without loss of generality, we may assume that ϕ is an everywhere-defined Borel function and $\phi \geq 0$.

Take $\delta > 0$ and let $\varepsilon > 0$ and $\mathcal{O}$ be as in (3.10) and (3.11). Let

$$\chi = \frac{\chi_{\mathcal{O}}}{m_G(\mathcal{O})}$$

and let $F : Y \to \mathbb{R}$ be defined as in (6.3). Then F is an everywhere-defined Borel function with $F \in L^p(\nu)$ and

$$\int_Y F \, d\nu = \int_X \phi \, d\mu.$$

Recall that it follows from Corollary 5.23 that the pointwise ergodic theorem holds for the family gG_t in $L^\infty \subset L^p$ for every $g \in G$. Using also the assumption of Theorem 6.10 and the invariance principle of Theorem 5.22,

$$\pi_Y(\beta_t)F(y) \to \int_Y \phi \, d\nu \quad \text{as } t \to \infty \tag{6.11}$$

for y in a G-invariant Lebesgue subset of Y of full measure. Then it follows from Lemma 6.1 that (6.11) holds for $y = \iota(e, x) = (e, x)\Gamma$ for x in a Γ-invariant Lebesgue subset of X of full measure. Arguing as in the proof of Proposition 2.1 in [GW], one shows that for every such x,

$$\limsup_{t \to \infty} \pi_X(\lambda_t)\phi(x) \leq (1 + \delta) \int_X \phi \, d\mu,$$

$$\liminf_{t \to \infty} \pi_X(\lambda_t)\phi(x) \geq (1 + \delta)^{-1} \int_X \phi \, d\mu,$$

for every $\delta > 0$. This completes the proof of Theorem 6.10. $\qquad\square$

6.6 EXPONENTIAL MEAN ERGODIC THEOREM

We now turn to proving the exponential mean ergodic theorem, and we start with the necessary prerequisite of establishing a quantitative error estimate in the lattice point–counting problem.

In this section we will assume that the family $\{G_t\}_{t>0}$ is admissible w.r.t. a family $\mathcal{O}_\varepsilon$ of finite upper local dimension ϱ_0. By definition, for $\varrho > \varrho_0$ and small $\varepsilon > 0$,

$$m_G(\mathcal{O}_\varepsilon) \geq C_\rho \varepsilon^\varrho. \tag{6.12}$$

Lemma 6.11. Effective error estimate of lattice points. *If the exponential mean ergodic theorem holds for the admissible family β_t acting in $(L^p(m_{G/\Gamma}), L^r(m_{G/\Gamma}))$ for some $p \geq r \geq 1$ with parameter $\theta_{p,r} > 0$, then*

$$\frac{|\Gamma_t|}{m_G(G_t)} = 1 + O(e^{-\alpha t}), \text{ where } \alpha = \frac{\theta_{p,r}}{\varrho(1 + r - \frac{r}{p}) + r}.$$

When the estimate $\left\|\pi^0_{G/\Gamma}(\beta_t)\right\|_{L^2_0 \to L^2_0} \leq C e^{-\theta t}$ holds, we can take $\theta_{p,r} = r\theta$.

Proof. As usual, let

$$\chi_\varepsilon = \frac{\chi_{\mathcal{O}_\varepsilon}}{m_G(\mathcal{O}_\varepsilon)}$$

and

$$\phi_\varepsilon(g\Gamma) = \sum_{\gamma \in \Gamma} \chi_\varepsilon(g\gamma),$$

so that that ϕ_ε is a bounded function on G/Γ with compact support,

$$\int_G \chi_\varepsilon \, dm_G = 1, \quad \text{and} \quad \int_{G/\Gamma} \phi_\varepsilon \, dm_{G/\Gamma} = 1.$$

It follows from the norm estimate given by the (L^p, L^r)-exponential mean ergodic theorem for β_t acting on G/Γ that for some fixed $\theta_{p,r} > 0$ and $C > 0$, and for every $\delta > 0, t > t_0$, and $\varepsilon > 0$,

$$m_{G/\Gamma}(\{x \in G/\Gamma : |\pi_{G/\Gamma}(\beta_t)\phi_\varepsilon(x) - 1| > \delta\}) \leq C\delta^{-r}\|\phi_\varepsilon\|^r_{L^p(G/\Gamma)}e^{-\theta_{p,r}t}.$$

By Lemma 6.2(1), for sufficiently small $\varepsilon > 0$,

$$\|\phi_\varepsilon\|_{L^p(G/\Gamma)} = m_G(\mathcal{O}_\varepsilon)^{1/p-1}.$$

We will choose both of the parameters ε and δ as a function of t and begin by requiring that the following condition holds:

$$C\delta^{-r} m_G(\mathcal{O}_\varepsilon)^{r/p-r} e^{-\theta_{p,r}t} = \frac{1}{2}m_G(\mathcal{O}_\varepsilon). \tag{6.13}$$

Then for sufficiently large t,

$$m_{G/\Gamma}(\{x \in G/\Gamma : |\pi_{G/\Gamma}(\beta_t)\phi_\varepsilon(x) - 1| > \delta\}) \leq \frac{1}{2}m_G(\mathcal{O}_\varepsilon).$$

Now as soon as $\mathcal{O}_\varepsilon$ maps injectively into G/Γ, we have

$$m_{G/\Gamma}(\mathcal{O}_\varepsilon\Gamma) = m_G(\mathcal{O}_\varepsilon),$$

and so we deduce that for every sufficiently large t, there exist $g_t \in \mathcal{O}_\varepsilon$ such that

$$|\pi_{G/\Gamma}(\beta_t)\phi_\varepsilon(g_t\Gamma) - 1| \leq \delta.$$

Then, using the claim from Lemma 6.7 and (3.6), we have for sufficiently large t,

$$|\Gamma_t| \leq (1 + \delta)m_G(G_{t+c\varepsilon}) \leq (1 + \delta)(1 + c\varepsilon)m_G(G_t),$$

provided only that ε, δ, and t satisfy condition (6.13).

In order to balance the two significant parts of the error estimate, let us take $c\varepsilon = \delta$. Then (6.13) together with (6.12) yields

$$C'e^{-\theta_{p,r}t} = \varepsilon^r m_G(\mathcal{O}_\varepsilon)^{1+r-r/p} \geq C''\varepsilon^{\varrho(1+r-r/p)+r}.$$

Thus we take both ε and δ to be constant multiples of

$$\exp\left(-\frac{\theta_{p,r}t}{\varrho(1+r-\frac{r}{p})+r}\right)$$

and conclude that

$$\left|\frac{|\Gamma_t|}{m_G(G_t)} - 1\right| \leq B\exp\left(-\frac{\theta_{p,r}t}{\varrho(1+r-\frac{r}{p})+r}\right),$$

where B is independent of t.

The lower estimate is proved similarly.

The last statement of the theorem follows immediately from Riesz-Thorin interpolation. $\qquad\square$

We now turn to the mean ergodic theorem with exponential rate of convergence.

Theorem 6.12. *Let β_t be an admissible family, let $\mathcal{O}_\varepsilon$ be of finite upper local dimension, and let $p \geq r > \varrho_0$. If the exponential mean ergodic theorem holds for β_t in $(L^p(v), L^r(v))$, then the exponential mean ergodic theorem holds for λ_t in $(L^p(\mu), L^r(\mu))$.*

Proof. We need to show that for some $\zeta = \zeta_{p,r} > 0$, $C = C_{p,r} > 0$, and every $\phi \in L^p(\mu)$,

$$\left\|\pi_X(\lambda_t)\phi - \int_X \phi\,d\mu\right\|_{L^r(\mu)} \leq Ce^{-\zeta t}\|\phi\|_{L^p(\mu)}.$$

Without loss of generality, we may assume that $\phi \geq 0$. As usual, let

$$\chi_\varepsilon = \frac{\chi\mathcal{O}_\varepsilon}{m_G(\mathcal{O}_\varepsilon)}$$

and let $F_\varepsilon : Y \to \mathbb{R}$ be defined as using χ_ε and ϕ in (6.3). Then $F_\varepsilon \in L^p(v)$ and

$$\int_Y F_\varepsilon\,dv = \int_X \phi\,d\mu.$$

By assumption, β_t satisfies the exponential mean ergodic theorem, so that for F_ε we have

$$\left\|\pi_Y(\beta_t)F_\varepsilon - \int_Y F_\varepsilon\,dv\right\|_{L^r(v)} \leq Ce^{-\theta t}\|F_\varepsilon\|_{L^p(v)} \tag{6.14}$$

for some $\theta = \theta_{p,r} > 0$.

Repeating the arguments in steps 1 and 2 of the proof of Theorem 6.8, we conclude that

(a) For sufficiently large t and every $(g, x) \in \mathcal{O}_\varepsilon \times X$,

$$\eta^{-1} \pi_Y(\beta_{t-c\varepsilon}) F_\varepsilon(\iota(g, x)) \le \pi_X(\lambda_t)\phi(x) \le \eta \pi_Y(\beta_{t+c\varepsilon}) F_\varepsilon(\iota(g, x)),$$

where

$$\eta = (1 + c\varepsilon)(1 + O(e^{-\alpha t})),$$

with α the error estimate in the lattice point count from Lemma 6.11.

(b) For sufficiently large t and small $\varepsilon > 0$,

$$\|\pi_Y(\beta_{t+c\varepsilon}) F_\varepsilon - \pi_Y(\beta_t) F_\varepsilon\|_{L^r(m_G \otimes \mu|_{\mathcal{O}_\varepsilon \times X})} \ll e^{-\theta t} \|F_\varepsilon\|_{L^p(\nu)}$$

and

$$\|\pi_Y(\beta_t) F_\varepsilon\|_{L^r(m_G \otimes \mu|_{\mathcal{O}_\varepsilon \times X})} \ll e^{-\theta t} \|F_\varepsilon\|_{L^p(\nu)} + \|\phi\|_{L^1(\mu)}.$$

Using (a), (b), and (6.14), we deduce as in the proof of Theorem 6.8 that

$$\left\| \pi_X(\lambda_t)\phi - \int_X \phi \, d\mu \right\|_{L^r(\mu)}$$
$$\le C' m_G(\mathcal{O}_\varepsilon)^{-1/r} \left(e^{-\theta t} \|F_\varepsilon\|_{L^p(\nu)} + (\varepsilon + e^{-\alpha t}) \|\phi\|_{L^1(\mu)} \right)$$

for every small $\varepsilon > 0$.

Using Lemma 6.2(1) and (2), we can estimate

$$\|F_\varepsilon\|_{L^p(\nu)} \le c_{p,\varepsilon} m_G(\mathcal{O}_\varepsilon)^{1/p-1} \|\phi\|_{L^p(\mu)}.$$

Now let ϱ be such that $\varrho_0 < \varrho < r$. By Hölder's inequality, $\|\phi\|_{L^1(\mu)} \le \|\phi\|_{L^p(\mu)}$, and collecting terms, we obtain

$$\left\| \pi_X(\lambda_t)\phi - \int_X \phi \, d\mu \right\|_{L^r(\mu)}$$
$$\le C'' m_G(\mathcal{O}_\varepsilon)^{-1/r} \left(e^{-\theta t} m_G(\mathcal{O}_\varepsilon)^{1/p-1} + \varepsilon + e^{-\alpha t} \right) \|\phi\|_{L^p(\mu)}$$
$$\le (e^{-\theta t} \varepsilon^{-\varrho(1+1/r-1/p)} + \varepsilon^{1-\varrho/r} + e^{-\alpha t} \varepsilon^{-\varrho/r}) \|\phi\|_{L^p(\mu)}.$$

Setting $\varepsilon = e^{-\delta t}$, with sufficiently small $\delta > 0$, we conclude the proof of Theorem 6.12. □

6.7 EXPONENTIAL STRONG MAXIMAL INEQUALITY

We now prove that the exponential decay of the norms $\pi_Y^0(\beta_t)$ in $L_0^2(Y)$, together with the ordinary strong maximal inequality for $\pi_Y(\beta_t)$ in some $L^q(Y)$, implies the exponential strong maximal inequality for λ_t following the method developed in [MNS] and [N4].

Theorem 6.13. *Let β_t be admissible averages w.r.t. $\mathcal{O}_\varepsilon$ of finite upper local dimension ϱ_0, satisfying*

- *the exponential mean ergodic theorem in $(L^p(v), L^r(v))$ with $\varrho_0 < r \le p$,*

- *the strong maximal inequality in $L^q(v)$ for some $q > 1$.*

Then λ_t satisfies the exponential strong maximal inequality in $(L^v(\mu), L^w(\mu))$ for v, w such that $1/v = (1-u)/q$ and $1/w = (1-u)/q + u/r$ for some $u \in (0, 1)$.

Proof. By Theorem 6.12, $\pi_X(\lambda_t)$ satisfies the exponential mean ergodic theorem: for every zero-integral function $f \in L_0^p(\mu)$ and $\zeta = \zeta_{p,r} > 0$,

$$\|\pi_X(\lambda_t)f\|_{L^r(\mu)} \le Ce^{-\zeta t}\|f\|_{L^p(\mu)}. \tag{6.15}$$

Consider an increasing sequence $\{t_i\}$ that contains all positive integers and divides each interval of the form $[n, n+1]$, $n \in \mathbb{N}$, into $\lfloor e^{r\zeta n/4} \rfloor$ subintervals of equal length. Then

$$t_{i+1} - t_i \le e^{-r\zeta \lfloor t_i \rfloor/4} \tag{6.16}$$

and

$$\sum_{i\ge 0} e^{-r\zeta t_i/2} \le \sum_{n\ge 0} \lfloor e^{r\zeta n/4} \rfloor e^{-r\zeta n/2} < \infty.$$

Hence, it follows from (6.15) that

$$\int_X \left(\sum_{i\ge 0} e^{r\zeta t_i/2} |\pi_X(\lambda_{t_i})f(x)|^r \right) d\mu(x) \le C'\|f\|_{L^p(\mu)}^r.$$

Setting

$$B_r(x, f) \stackrel{\text{def}}{=} \left(\sum_{i\ge 0} e^{r\zeta t_i/2} |\pi_X(\lambda_{t_i})f(x)|^r \right)^{1/r},$$

we have, for some C'' independent of f,

$$|\pi_X(\lambda_{t_i})f(x)| \le B_r(x, f)e^{-\zeta t_i/2}, \tag{6.17}$$

$$\|B_r(\cdot, f)\|_{L^r(\mu)} \le C''\|f\|_{L^p(v)}. \tag{6.18}$$

Claim. *For any sufficiently large t and t_i such that $|t - t_i| \ll e^{-r\zeta \lfloor t \rfloor/4}$, we have for every $f \in L^\infty(\mu)$,*

$$|\pi_X(\lambda_t)f - \pi_X(\lambda_{t_i})f| \le e^{-\eta t}\|f\|_{L^\infty(\mu)},$$

with some $\eta > 0$, independent of t and given explicitly below.

Proof. To prove the claim, note that it follows from (6.16) that t_i satisfying the first property exists. Without loss of generality we suppose that $t > t_i$. Then,

$$\pi_X(\lambda_t)f(x) - \pi_X(\lambda_{t_i})f(x)$$

$$= \frac{1}{|\Gamma_t| \cdot |\Gamma_{t_i}|} \left(|\Gamma_{t_i}| \sum_{\gamma \in \Gamma_t} f(\gamma^{-1} \cdot x) - |\Gamma_t| \sum_{\gamma \in \Gamma_{t_i}} f(\gamma^{-1} \cdot x) \right)$$

$$= -\frac{|\Gamma_t - \Gamma_{t_i}|}{|\Gamma_t| \cdot |\Gamma_{t_i}|} \sum_{\gamma \in \Gamma_{t_i}} f(\gamma^{-1} \cdot x) + \frac{1}{|\Gamma_t|} \sum_{\gamma \in \Gamma_t - \Gamma_{t_i}} f(\gamma^{-1} \cdot x)$$

$$\le \frac{2|\Gamma_t - \Gamma_{t_i}|}{|\Gamma_t|} \|f\|_{L^\infty(\mu)}.$$

Applying the estimate provided by Lemma 6.11 and (3.6), we obtain the estimate

$$\frac{|\Gamma_t - \Gamma_{t_i}|}{|\Gamma_t|} = 1 - \frac{|\Gamma_{t_i}|}{|\Gamma_t|} = 1 - \frac{1 + O(e^{-\alpha t_i})}{1 + O(e^{-\alpha t})} \cdot \frac{m_G(G_{t_i})}{m_G(G_t)}$$

$$\leq 1 - (1 + O(e^{-\alpha t})) \cdot \frac{1}{1 + c(t - t_i)}.$$

This estimate implies the claim.   □

Continuing with the proof of Theorem 6.13, we use (6.17) and the claim and deduce that for $\delta = \min\{\zeta/2, \eta\}$ and every $f \in L_0^\infty(\mu)$,

$$|\pi_X(\lambda_t)f(x)| \leq |\pi_X(\lambda_{t_i})f(x)| + |\pi_X(\lambda_t)f(x) - \pi_X(\lambda_{t_i})f(x)|$$

$$= (B_r(x, f) + \|f\|_{L^\infty(\mu)}) e^{-\delta t}.$$

Hence, by (6.17) and (6.18), for some $t_0, C > 0$ and every $f \in L_0^\infty(\mu)$,

$$\left\| \sup_{t \geq t_0} e^{\delta t} |\pi_X(\lambda_t)f| \right\|_{L^r(\mu)} \leq C\|f\|_{L^\infty(\mu)}, \tag{6.19}$$

where we have used the fact that since μ is a probability measure, for $f \in L^\infty(\mu)$, $\|f\|_{L^r(\mu)} \leq \|f\|_{L^\infty(\mu)}$.

Now for a measurable function $\tau : X \to [t_0, \infty)$ and $z \in \mathbb{C}$, we consider the linear operator (acting on functions $f : X \to \mathbb{C}$)

$$U_z^\tau f(x) = e^{z\delta\tau(x)} \left(\pi_X(\lambda_{\tau(x)})f(x) - \int_X f \, d\mu \right).$$

By the strong maximal inequality for λ_t in $L^q(\mu)$ (which holds using Theorem 6.3(1) and our second assumption), when $\operatorname{Re} z = 0$, the operator

$$U_z^\tau : L^q(\mu) \to L^q(\mu)$$

is bounded. By (6.19), when $\operatorname{Re} z = 1$, the operator

$$U_z^\tau : L^\infty(\mu) \to L^r(\mu)$$

is also bounded, with bounds independent of the function τ. Hence, by the complex interpolation theorem (see, e.g., [MNS] for a fuller discussion) for every $u \in (0, 1)$ and v, w such that

$$1/v = (1 - u)/q \quad \text{and} \quad 1/w = (1 - u)/q + u/r,$$

we have

$$\left\| \sup_{t \geq t_0} e^{u\delta t} \left| \pi_X(\lambda_t)f - \int_X f \, d\mu \right| \right\|_{L^w(\mu)} \leq C\|f\|_{L^v(\mu)}.$$

This completes the proof of Theorem 6.13.   □

6.8 COMPLETION OF THE PROOFS

Completion of the proof of Theorem 6.3. Note that parts 1–5 of Theorem 6.3 are proved in Theorems 6.6, 6.8, 6.10, 6.12, and 6.13, respectively. Part 6, the exponential pointwise ergodic theorem, is a consequence of part 5, the exponential strong maximal inequality.   ☐

Completion of the proof of Theorem 6.4. Note that parts 1–4 of Theorem 6.4 are proved in Lemmas 6.5, 6.7, 6.10, and 6.11, respectively.   ☐

Completion of the proof of Theorem 4.5. Clearly, parts 1–3 of Theorem 6.3 together imply Theorem 4.5, provided only that the admissible averages β_t do indeed satisfy the mean, maximal, and pointwise ergodic theorems in $L^p(\nu)$ (and as a result also in $L^p(m_{G/\Gamma})$). This follows immediately from Theorem 4.1, also taking into account the fact that since we are considering the action induced to G^+, the action is necessarily totally weak-mixing since G^+ has no nontrivial finite-dimensional representations.   ☐

Completion of the proof of Theorem 4.6. Since the formulation of Theorem 6.3(4) incorporates the assumption that $p \geq r > \varrho_0$, where ϱ_0 is the upper local dimension, in order to complete the proof of Theorem 4.6, we must remove this restriction.

Let us consider the exponential mean ergodic theorem in (L^p, L^r) first. By Theorem 4.2, β_t on G satisfies this theorem for all $p = r > 1$, and hence by Theorem 6.12, we obtain that λ_t satisfies it if $p > \varrho$. But clearly, $\lambda_t - \int_X d\mu$ has norm bounded by 2 in every L^p, $1 \leq p \leq \infty$. By Riesz-Thorin interpolation, it follows that λ_t satisfies the exponential mean ergodic theorem in every L^p, $1 < p < \infty$, and hence in (L^p, L^r), $p \geq r \geq 1$, $(p, r) \neq (1, 1)$ or (∞, ∞).

As to the exponential strong maximal inequality, note first that by Theorem 4.2 β_t satisfies the (L^∞, L^2) exponential strong maximal inequality in every action of G where $\|\pi_Y(\beta_t)\|_{L^2} \leq C \exp(-\theta t)$. It follows that β_t satisfies the exponential strong maximal inequality in (L^∞, L^r) for a finite $r > \varrho_0 \geq 1$. By Theorem 5.13, β_t also satisfies the standard strong maximal inequality in every L^q, $q > 1$. Thus by Theorem 6.13, the exponential strong maximal inequality in (L^v, L^w) holds for the averages λ_t (provided the norm exponential decay condition holds in the induced action, which is the case under our assumptions). By their explicit formula it is clear that we can choose v to be as close as we like to 1, thus determining a consequent $w < v$ and some positive rate of exponential decay.

This completes the proof of Theorem 4.6.   ☐

Completion of the proofs of the ergodic theorems for connected semisimple Lie groups and their lattices. The ergodic theorems for connected semisimple Lie groups and their lattices stated in §1.2–1.4 all follow from Theorems 4.1, 4.2, 4.5, 4.6, and 4.8, together with Lemma 6.11 (or more precisely Corollary 8.1 below). This is a straightforward verification, bearing in mind that every nontrivial unitary representation of a connected semisimple Lie group with finite center is totally weak-mixing.   ☐

Completion of the proofs of Theorems 1.12 and 4.8. The first part of these two results asserts that β_t and λ_t defined using the $CAT(0)$-metric satisfy the pointwise ergodic action in any ergodic action of the group or the lattice even in the reducible case. Theorem 4.5 establishes this result based on Theorem 4.1 since the averages in question are indeed admissible, well balanced, and boundary-regular. In the Riemannian case this was established in [MNS], and in general it is established in Chapter 7 below in the proof of Theorem 3.18. It follows that an exponential decay estimate holds for the "sphere" averages $\|\pi_Y(\partial \beta_t)\|$ even in the case of a reducible action, and thus pointwise convergence on a dense subspace holds. The last part, equidistribution in isometric lattice actions, is proved in §6.9 below.

In might be worth commenting that for the *sequences* of averages β_n and λ_n the entire Sobolev space argument is of course superfluous. Since the strong maximal inequality holds, as well as pointwise convergence on dense subspace, it follows that the pointwise ergodic theorem holds for both sequences, even in the reducible case. □

This concludes the proofs of the ergodic theorems for actions of the group or the lattice.

We now turn to a discussion of equidistribution.

6.9 EQUIDISTRIBUTION IN ISOMETRIC ACTIONS

Let us prove the following generalization of Theorem 1.9.

Theorem 6.14. *Let G be an S-algebraic group as in Definition 3.4 over fields of characteristic zero. Let $\Gamma \subset G^+$ be a lattice and $G_t \subset G^+$ an admissible one-parameter family or sequence. Let (S, m) be an isometric action of Γ on a compact metric space S, preserving an ergodic probability measure m of full support. If Γ is an irreducible lattice, then*

$$\lim_{n \to \infty} \max_{s \in S} \left| \pi_S(\lambda_t) f(s) - \int_S f \, dm \right| = 0$$

and, in particular, $\pi_S(\lambda_t) f(s) \to \int_S f \, dm$ for every $s \in S$.

For any S-algebraic G and lattice $\Gamma \subset G^+$, the same results hold provided that G_t are left-radial and balanced.

Under these conditions, the action is uniquely ergodic.

Proof. When G is defined over fields of characteristic zero, an action of G induced by an isometric ergodic action of an irreducible lattice is an irreducible action of G, as shown in [St]. Then, according to Theorem 4.5, λ_t satisfies the mean ergodic theorem in $L^2(S, m)$. The mean ergodic theorem also holds for general G and Γ, provided only that G_t is balanced and left-radial, as asserted in Theorem 4.5. The proof is complete upon using the next result. □

To complete the proof it suffices to show that the mean ergodic theorem in an isometric action implies equidistribution and unique ergodicity. We now turn to this result.

Proposition 6.15. *Let the discrete group Γ act isometrically on a compact metric space (S, m), preserving an ergodic probability measure of full support. If a one-parameter family (or sequence) of averages λ_t on Γ satisfy the mean ergodic theorem in $L^2(S, m)$, then $\pi_S(\lambda_t)f$ converges uniformly to the constant $\int_S f\,dm$ for every continuous $f \in C(S)$, and the action is uniquely ergodic.*

Proof. Our proof is a straightforward generalization of [G, Thm. 2] (where the case of free groups is considered) and is brought here for completeness.

Given a function $f \in C(S)$, consider the set $C(f)$ consisting of f together with $\pi_S(\lambda_t)f$, $t \in \mathbb{R}_+$. $C(f)$ constitutes an equicontinuous family of functions since Γ acts isometrically on S. Thus $C(f)$ has compact closure in $C(S)$ w.r.t. the uniform norm. Let $f_0 \in C(S)$ be the uniform limit of $\pi_S(\lambda_{t_i})f$ for some subsequence $t_i \to \infty$. Then f_0 is of course also the limit of $\pi_S(\lambda_{t_i})f$ in the $L^2(S, m)$-norm. Given that λ_t satisfy the mean ergodic theorem, it follows that $f_0(s) = \int_S f\,dm$ for m–almost all $s \in S$. Since m has full support, the last equality holds on a dense subset of S, and since f_0 is continuous, it must hold everywhere. Thus $\pi_S(\lambda_{t_i})f$ converges uniformly to the constant $\int_S f\,dm$, and this holds for every subsequence $t_i \to \infty$. It follows that the latter constant is the unique limit point in the uniform closure of the family $\pi_S(\lambda_t)f$. Hence

$$\lim_{t \to \infty} \left\| \pi_S(\lambda_t)f - \int_S f\,dm \right\|_{C(S)} = 0.$$

Unique ergodicity follows from the fact that for any ergodic Γ-invariant measure m' and any continuous function f, $\pi_S(\lambda_t)f(s)$ converges to $\int_S f\,dm$ for every $s \in S$ but also converges to $\int_S f\,dm'$ in $L^2(S, m')$. Hence the two integrals must be equal for any continuous f, so that the measures m and m' are equal. $\qquad\square$

Chapter Seven

Volume estimates and volume regularity

The present chapter is devoted to establishing regularity properties of the standard averages and more general ones, as well as to establishing conditions sufficient for the averages to be balanced or well balanced. We will also discuss boundary-regularity and differentiability properties of volume functions for some metrics, particularly, $CAT(0)$-metrics.

7.1 ADMISSIBILITY OF STANDARD AVERAGES

We begin with a proof of Theorem 3.15, whose statement we recall.

Theorem 3.15. *For an S-algebraic group $G = G(1) \cdots G(N)$ as in Definition 3.4, the following families of sets $G_t \subset G$ are admissible, where a_i are any positive constants, and $t \in \mathbb{R}_+$ when S contains at least one infinite place and $t \in \mathbb{N}_+$ otherwise.*

1. *Let S consist of infinite places, and let $G(i)$ be a closed subgroup of the isometry group of a symmetric space $X(i)$ of nonpositive curvature equipped with the Cartan-Killing metric. Let d_i be the associated distance, and for $u_i, v_i \in X(i)$, define*

$$G_t = \{(g_1, \ldots, g_N) : \sum_i a_i d_i(u_i, g_i \cdot v_i) < t\}.$$

2. *Let S consist of infinite places and let $\rho_i : G(i) \to \mathrm{GL}(V_i)$ be proper rational representations. For norms $\| \cdot \|_i$ on $\mathrm{End}(V_i)$, define*

$$G_t = \{(g_1, \ldots, g_N) : \sum_i a_i \log \|\rho_i(g_i)\|_i < t\}.$$

3. *For infinite places, let $X(i)$ be the symmetric space of $G(i)$ equipped with the Cartan-Killing distance d_i, and for finite places, let $X(i)$ be the Bruhat-Tits building of $G(i)$ equipped with the path metric d_i on its 1-skeleton. For $u_i \in X(i)$, define*

$$G_t = \{(g_1, \ldots, g_N) : \sum_i a_i d_i(u_i, g_i \cdot u_i) < t\}.$$

4. *Let $\rho_i : G(i) \to \mathrm{GL}(V_i)$ be proper representations, rational over the fields of definition F_i. For infinite places, let $\| \cdot \|_i$ be a Euclidean norm on $\mathrm{End}(V_i)$*

and assume that $\rho_i(G(i))$ is self-adjoint: $\rho_i(G(i))^t = \rho_i(G(i))$. For finite places, let $\| \cdot \|_i$ be the max-*norm on* End(V_i). *Define*

$$G_t = \{(g_1, \ldots, g_N) : \sum_i a_i \log \|\rho_i(g_i)\|_i < t\}.$$

The proof is divided into several propositions. To handle Archimedean groups we will employ some arguments originating in [DRS] and [EMS], and in the general case of S-algebraic groups, we will also employ convolution arguments which will be developed below. We note that the latter arguments will utilize knowledge of the behavior of vol(G_t) for *all* $t > 0$, and we will thus consider below the behavior for t large and for t small, separately. In order to establish regularity properties of the volume function vol(G_t), we will use a Tauberian-type argument (as in [CT]) and establish meromorphic continuation of the Mellin transform for the volume function with help of the Hironaka resolution of singularities. This idea goes back to Atiyah [A] and Bernstein and Gelfand [BeGe]. We also refer the reader to [BO] for another argument establishing volume regularity in a particular case.

Proposition 7.1. *Let G be a connected semisimple group with finite center and X the corresponding symmetric space equipped with the Cartan-Killing metric d. Let d be the associated distance, and for $u \in X$, set*

$$G_t = \{g \in G; \, d(u, g \cdot u) < t\}.$$

Then there exists $c > 0$ such that for all $t \geq 0$ and $\epsilon \in (0, 1)$,

$$\mathrm{vol}(G_{t+\epsilon}) - \mathrm{vol}(G_t) \leq c\epsilon \max\{1, \mathrm{vol}(G_t)\}.$$

Proof. Note that the stabilizer of u is a maximal compact subgroup K of G, and for a Cartan subgroup A of G, the map $a \mapsto a \cdot u, a \in A$, is an isometry. We introduce polar coordinates $(r, \omega) \in \mathbb{R}^+ \times S^+$ on the Lie algebra of A, where S^+ is the intersection of the unit sphere in $\mathfrak{a}$ with the positive Weyl chamber. With respect to the Cartan decomposition $G = KA^+K$, the Haar measure on G is given by $\xi(r, \omega) \, dr d\omega dk$ with a nonnegative continuous density function ξ. Then

$$\mathrm{vol}(G_{t+\epsilon}) - \mathrm{vol}(G_t) = \int_{S^+} \int_t^{t+\epsilon} \xi(r, \omega) \, dr d\omega = \epsilon \int_{S^+} \xi(\sigma(\omega), \omega) \, d\omega$$

for some $\sigma(\omega) \in [t, t + \epsilon]$. This implies the claim for $t \in [0, 1]$. To establish the claim for $t > 1$, we use the following property of the function ξ (see [EMS, Lem. A.3]): there exists $c > 0$ such that for every $r > 1$ and $\omega \in S^+$,

$$\xi(r, \omega) \leq c \int_0^r \xi(s, \omega) ds. \tag{7.1}$$

Substituting into the previous inequality completes the proof. $\square$

Proposition 7.2. *Let d be the Cartan-Killing metric on a symmetric space X of nonpositive curvature, $u, v \in X$, and G a closed connected semisimple subgroup of the isometry group of X. Define the sets*

$$G_t = \{g \in G; \, d(u, g \cdot v) < t\}.$$

Then there exist $c, t_0 > 0$ such that for all $t > t_0$ and $\epsilon \in (0, 1)$,

$$\mathrm{vol}(G_{t+\epsilon}) - \mathrm{vol}(G_t) \leq c\epsilon \, \mathrm{vol}(G_t).$$

Proof. As already noted, there exist a maximal compact subgroup K of G and an associated Cartan subgroup A such that $K \subset \mathrm{Stab}(v)$ and the map $a \mapsto a \cdot v$, $a \in A$, is an isometry. Consider polar coordinates $(r, \omega) \in \mathbb{R}^+ \times S^+$ on the Lie algebra of A and set

$$S_t(k) = \{(r, \omega); \; d(k^{-1}u, \exp(r\omega)v) < t\}.$$

Then

$$\mathrm{vol}(G_t) = \int_K \int_{S_t(k)} \xi(r, \omega) \, dr \, d\omega \, dk.$$

For $p \in Ku$, we consider the function

$$f_p(x) = \frac{1}{2} d(p, x)^2, \quad x \in X.$$

We have

$$\mathrm{grad} \, f_p = -\exp_x^{-1}(p),$$

and for the unit-speed geodesic ray $\gamma_\omega(r) = \exp(r\omega)v$,

$$\frac{d}{dr} f_p(\gamma_\omega(r)) = \langle (\mathrm{grad} \, f_p)_{\gamma_\omega(r)}, \gamma'_\omega(r) \rangle_{\gamma_\omega(r)},$$

$$\frac{d^2}{dr^2} f_p(\gamma_\omega(r)) = \langle \nabla_{\gamma'_\omega(r)} (\mathrm{grad} \, f_p)_{\gamma_\omega(r)}, \gamma'_\omega(r) \rangle_{\gamma_\omega(r)}.$$

Since the space X has nonpositive sectional curvature, we deduce (see [J, Thm. 4.6.1]) that

$$\langle \nabla_w (\mathrm{grad} \, f_p)_x, w \rangle_x \geq \|w\|_x^2 \quad \text{for every } w \in T_x X.$$

Hence,

$$\frac{d^2}{dr^2} f_p(\gamma_\omega(r)) \geq 1. \tag{7.2}$$

Therefore, there exist $r_0 > 0$ and $\alpha > 0$ such that for every $r > r_0$, $p \in Ku$, and $\omega \in S^+$, we have

$$\frac{d}{dr} f_p(\gamma_\omega(r)) \geq \alpha r.$$

For $\epsilon > 0$ and $r > r_0$,

$$d(p, \exp((r + \epsilon)\omega)v) = \sqrt{2 f_p(\gamma_\omega(r + \epsilon))} \geq \sqrt{2 f_p(\gamma_\omega(r)) + 2\alpha r \epsilon}$$

$$= d(p, \exp(r\omega)v) \sqrt{1 + 2\alpha r \epsilon / d(p, \gamma_\omega(r))^2}.$$

Since it follows from the triangle inequality that, for some $c > 0$,

$$r - c \leq d(p, \gamma_\omega(r)) \leq r + c \quad \text{for all } p \in Ku \text{ and } \omega \in S^+,$$

we conclude that, for some $\beta > 0$,

$$d(p, \exp((r + \epsilon)\omega)v) \geq d(p, \exp(r\omega)v) + \beta \epsilon \tag{7.3}$$

for sufficiently large r and sufficiently small $\epsilon > 0$. This implies that for sufficiently large t, the sets $S_t(k)$ are star-shaped. Let $r_t(k, \omega)$ be the unique solution of the equation

$$d(k^{-1}u, \exp(r\omega)v) = t.$$

Note that, by (7.3), for sufficiently large t and $\epsilon \in (0, 1)$,

$$r_{t+\epsilon}(k, \omega) \leq r_t(k, \omega) + \beta^{-1}\epsilon.$$

Setting

$$m_t(k, \omega) = \int_0^{r_t(k,\omega)} \xi(r, \omega)dr,$$

we have

$$m_{t+\epsilon}(k, \omega) - m_t(k, \omega) \leq \int_0^{r_t(k,\omega)+\beta^{-1}\epsilon} \xi(r, \omega)r\,dr - \int_0^{r_t(k,\omega)} \xi(r, \omega)dr$$

$$= \beta^{-1}\epsilon \cdot \xi(\sigma, \omega)$$

for some $\sigma \in [r_t(k, \omega), r_t(k, \omega) + \beta^{-1}\epsilon]$. Now it follows from (7.1) that

$$m_{t+\epsilon}(k, \omega) - m_t(k, \omega) \leq (\beta^{-1}c)\epsilon \cdot m_{t+\epsilon}(k, \omega).$$

This shows that

$$\text{vol}(G_{t+\epsilon}) = \int_K \int_{S^+} m_{t+\epsilon}(k, \omega)d\omega dk \leq (\beta^{-1}c)\epsilon \cdot \text{vol}(G_{t+\epsilon}) + \text{vol}(G_t),$$

which implies the proposition. $\qquad\square$

Proposition 7.3. (cf. [EMS, App.]). *Let $\rho : G \to \text{GL}(V)$ be a proper representation of a connected semisimple Lie group G, $\| \cdot \|$ a norm on $\text{End}(V)$, and*

$$G_t = \{g \in G;\ \log \|\rho(g)\| < t\}.$$

Then there exist $c, t_0 > 0$ such that for all $t > t_0$ and all $\epsilon \in (0, 1)$,

$$\text{vol}(G_{t+\epsilon}) - \text{vol}(G_t) \leq c\epsilon\ \text{vol}(G_t).$$

Proof. We employ the argument from [EMS, App.], but since this argument is not quite complete (see (7.4) below), we provide a sketch which indicates that in our setting it is indeed applicable and provides the Lipschitz estimate.

We fix a Cartan decomposition $G = K A^+ K$ and use polar coordinates (r, ω) on the Lie algebra of A. Let

$$S_t(k_1, k_2) = \{(r, \omega);\ \|\rho(k_1 \exp(r\omega)k_2)\| < e^t\}.$$

Then

$$\text{vol}(G_t) = \int_{K \times K} \int_{S_t(k_1, k_2)} \xi(r, \omega)dr d\omega dk_1 dk_2.$$

Since $\rho(A)$ is (simultaneously) diagonalizable over $\mathbb{R}$, there exist vectors $v_i \in \text{End}(\mathbb{R}^n)$ such that

$$\rho(\exp(r\omega)) = \sum_i e^{r\lambda_i(\omega)}v_i.$$

Since all norms are equivalent and $\rho(K)$ is compact, there exist $c_1, c_2 > 0$ such that for every $k_1, k_2 \in K$, $r > 0$, and $\omega \in S^+$,

$$c_1 \exp(r \max_i \lambda_i(\omega)) \leq \|\rho(k_1 \exp(r\omega)k_2)\| \leq c_2 \exp(r \max_i \lambda_i(\omega)). \qquad (7.4)$$

Note that the lower estimate needs a further argument in the generality of [EMS] since the v_i depend on ω. But for constant v_i, using these estimates, the argument from [EMS, Lem. A.4] shows that

1. There exists $t_0 > 0$ such that for $t > t_0$ the sets $S_t(k_1, k_2)$ are star-shaped.

2. There exists $r_0 > 0$ such that for every $r > r_0$, $\omega \in S^+$, and $\epsilon \in [0, 1)$, we have

$$\|\rho(k_1 \exp((r + \epsilon)\omega)k_2)\| \geq g(\epsilon) \cdot \|\rho(k_1 \exp(r\omega)k_2)\|,$$

   where $g : [0, 1) \to [1, \infty)$ is an explicit smooth function such that $g(0) = 1$ and $g' > 0$. In particular, there exists $\beta > 0$ such that $g(\epsilon) \geq e^{\beta\epsilon}$.

Let $r_t(k_1, k_2, \omega)$ denote the unique solution of the equation

$$\|\rho(k_1 \exp(r\omega)k_2)\| = e^t.$$

Then it follows that

$$r_{t+\epsilon}(k_1, k_2, \omega) \leq r_t(k_1, k_2, \omega) + \beta^{-1}\epsilon.$$

Finally, the Lipschitz property of the sets G_t can be proved as in Proposition 7.1 above. $\qquad \square$

Let us note the following consequence of the foregoing arguments. Let H be a symmetric subgroup of a connected semisimple Lie group with finite center. Namely, the Lie algebra $\mathfrak{h}$ is the fixed-point set of an involution of the Lie algebra $\mathfrak{g}$, and the homogeneous space G/H can be embedded as a Zariski closed G-orbit in a linear space V on which G acts via a linear representation.

Corollary 7.4. *Proposition 7.3 applies to the family B_t of subsets of an affine symmetric variety $G/H \subset V$ defined by an arbitrary norm on the ambient vector space V, namely, $B_t = \{gH \in G/H \; ; \; \log \|gv\| < t\}$, where $H = St_G(v)$.*

Proof. Indeed, for symmetric varieties one has a decomposition of the form $G = KAH$, where A is simultaneously diagonalizable, and the arguments utilized in the proof of Proposition 7.3 apply without any material changes. $\qquad \square$

Proposition 7.5. *Let $\rho : G \to GL(V)$ be a proper representation of a connected semisimple Lie group G such that ${}^t\rho(G) = \rho(G)$, $\| \cdot \|$ is the Euclidean norm on $End(V)$, and*

$$G_t = \{g \in G; \log \|\rho(g)\| < t\}.$$

Then there exists $c > 0$ such that for all $t \in \mathbb{R}$ and all $\epsilon \in (0, 1)$,

$$\text{vol}(G_{t+\epsilon}) - \text{vol}(G_t) \leq c\epsilon \, \max\{1, \text{vol}(G_t)\}.$$

Proof. Let t_0 be as in Proposition 7.3. It remains to prove the claim for $t \leq t_0$.

Since $\rho(G)$ is self-adjoint, there exist a maximal compact subgroup K such that $\rho(K) \subset SO(V)$ and a Cartan subgroup such that $\rho(A)$ is diagonal. For $k_1, k_2 \in K$ and $a \in \text{Lie}(A)$,

$$\|\rho(k_1 \exp(a)k_2)\|^2 = \sum_i e^{2\lambda_i(a)},$$

where λ_i are characters of $\text{Lie}(A)$ such that $\sum_i \lambda_i = 0$. We use polar coordinates (r, ω) on A and set

$$f_\omega(r) = \sum_i e^{2r\lambda_i(\omega)}.$$

Then

$$\text{vol}(G_t) = \int_{(r,\omega):\log f_\omega(r) < 2t} \xi(r, \omega) \, dr \, d\omega.$$

Since $f_\omega'' > 0$ and $f_\omega'(0) = 0$, the function $\log f_\omega$ is increasing. Let r_ω be the inverse function of $\log f_\omega$ and $r_0 = \max\{r_\omega(2t_0); \omega \in S^+\}$. By the mean value theorem, there exists $\alpha > 0$ such that

$$f_\omega'(r) \geq \alpha r \quad \text{for } r \in [0, r_0].$$

Then for some $\beta > 0$,

$$r_\omega'(t) \leq \beta r_\omega(t)^{-1} \quad \text{for } t \leq 2t_0. \tag{7.5}$$

Since for some $c > 0$,

$$\xi(r, \omega) \leq cr \quad \text{for } r \in [0, r_0] \text{ and } \omega \in S^+,$$

we have

$$\text{vol}(G_{t+\epsilon}) - \text{vol}(G_t) = \int_{S^+} \int_{r_\omega(2t)}^{r_\omega(2t+2\epsilon)} \xi(r, \omega) \, dr \, d\omega$$

$$\leq c/2 \int_{S^+} (r_\omega(2t + 2\epsilon)^2 - r_\omega(2t)^2) \, d\omega.$$

By the mean value theorem, the integrand is bounded by $2r_\omega(\theta)r_\omega'(\theta)$ for some $\theta \in [2t, 2t + 2\varepsilon]$. Hence, the proposition follows from (7.5). $\square$

7.2 CONVOLUTION ARGUMENTS

We now discuss convolution arguments, which together with the foregoing result will complete the proof of Theorem 3.15. Let $X(i)$, $i = 1, 2$, be locally compact spaces and let $d_i : X(i) \to [t_0, \infty)$, $i = 1, 2$, be proper continuous functions. We set

$$v_i(t) = \text{vol}(\{x_i \in X(i); d_i(x_i) < t\}),$$
$$v(t) = \text{vol}(\{(x_1, x_2) \in X(1) \times X(2); d_1(x_1) + d_2(x_2) < t\}).$$

Proposition 7.6. *Suppose that there exist $c > 0$ and $s_0 > t_0$ such that for all sufficiently small $\epsilon > 0$ and for all $t > s_0$,*

$$v_i(t + \epsilon) \leq (1 + c\epsilon)\, v_i(t), \quad i = 1, 2.$$

Then for all sufficiently small $\epsilon > 0$ and for all $t > 2s_0 + 2$,

$$v(t + \epsilon) \leq (1 + 3c\epsilon)\, v(t).$$

Proof. Let

$$w_1(t) := \text{vol}(\{(x_1, x_2); d_1(x_1) + d_2(x_2) < t, d_1(x_1) \geq s_0 + 1, d_2(x_2) < s_0 + 1\}),$$
$$w_2(t) := \text{vol}(\{(x_1, x_2); d_1(x_1) + d_2(x_2) < t, d_1(x_1) < s_0 + 1, d_2(x_2) \geq s_0 + 1\}),$$
$$w_3(t) := \text{vol}(\{(x_1, x_2); d_1(x_1) + d_2(x_2) < t, d_1(x_1) \geq s_0 + 1, d_2(x_2) \geq s_0 + 1\}).$$

For sufficiently small ϵ and for all $t \geq 2s_0 + 2$, we have

$$w_1(t + \epsilon) - w_1(t)$$

$$\leq \int_{x_2 : d_2(x_2) < s_0 + 1} \text{vol}(\{x_1 : \max\{s_0 + 1, t - d_2(x_2)\} \leq d_1(x_1)$$
$$< t + \epsilon - d_2(x_2)\})\, dx_2$$

$$\leq \int_{x_2 : t - d_2(x_2) > s_0} (v_1(t - d_2(x_2) + \epsilon) - v_1(t - d_2(x_2)))\, dx_2$$

$$\leq c\epsilon \int_{x_2 : t - d_2(x_2) > s_0} v_1(t - d_2(x_2))\, dx_2 \leq c\epsilon\, v(t).$$

Using a similar argument, we show that

$$w_i(t + \epsilon) - w_i(t) \leq c\epsilon\, v(t), \quad i = 1, 2, 3,$$

for sufficiently small ϵ and for all t. Since for $t > 2s_0 + 2$,

$$v(t + \epsilon) - v(t) = (w_1(t + \epsilon) - w_1(t)) + (w_2(t + \epsilon) - w_2(t)) + (w_3(t + \epsilon) - w_3(t)),$$

this implies the claim. $\square$

A very similar argument establishes the following, with v_i and v defined as above.

Proposition 7.7. *Suppose that there exist $c > 0$ and $s_0 > t_0$ such that for all $t > s_0$,*

$$v_i(t + 1) \leq c\, v_i(t), \quad i = 1, 2.$$

Then for all all $t > 2s_0 + 2$,

$$v(t + 1) \leq (1 + 3c)\, v(t).$$

We shall also need another version of Proposition 7.6.

Proposition 7.8. *Assume that $X(1)$ is noncompact. Suppose that there exists $c > 0$ such that for all sufficiently small $\epsilon > 0$ and for all $t \geq t_0$,*

$$v_1(t + \epsilon) - v_1(t) \leq c\epsilon \max\{v_1(t), 1\}.$$

Then there exists $s_0 \geq 0$ such that for all sufficiently small $\epsilon > 0$ and for all $t \geq 2t_0$,

$$v(t + \epsilon) - v(t) \leq c\epsilon\, v(t + s_0).$$

Proof. Since

$$v(t) = \int_{X(2)} v_1(t - d_2(x_2)) \, dx_2 = \int_{x_2 : t - d_2(x_2) \geq t_0} v_1(t - d_2(x_2)) \, dx_2,$$

it follows that for all sufficiently small $\epsilon > 0$ and $t \geq 2t_0$,

$$v(t + \epsilon) - v(t) \leq c\epsilon \int_{x_2 : t - d_2(x_2) \geq t_0} \max\{v_1(t - d_2(x_2)), 1\} \, dx_2.$$

Since $X(1)$ is noncompact, $v(t) \to \infty$ as $t \to \infty$ and there exists $s_0 > 0$ such that for all $t > s_0 + t_0$, we have $v_1(t) > 1$. Then

$$v(t + \epsilon) - v(t) \leq c\epsilon \int_{x_2 : t - d_2(x_2) \geq t_0} v_1(s_0 + t - d_2(x_2)) \, dx_2$$

$$\leq c\epsilon \, v(t + s_0).$$

This completes the proof. ☐

Proof of Theorem 3.15. It is clear that G_t is an admissible sequence when S consists of finite places. Now we assume that S contains at least one infinite place. Part 1 follows from Propositions 7.2 and 7.6. Part 2 follows from Propositions 7.3 and 7.6. Part 3 follows from Propositions 7.1, 7.7, and 7.8. Part 4 follows from Propositions 7.5, 7.7, and 7.8. ☐

As we saw, the properties of balancedness and well balancedness play an important role in the proofs of the ergodic theorems. To complete our discussion of the averages discussed in Theorem 3.15, let us note the following.

First, the following criterion is sufficient to establish that the averages defined by the metric $(\sum_i a_i d_i^p)^{1/p}$ are in fact well balanced, provided $1 < p < \infty$.

Proposition 7.9. *Suppose that for some $s_0 > t_0, a_i, b_i > 0, u_i \geq 0,$ and $w_i > 0,$*

$$a_i t^{u_i} e^{w_i t} \leq v_i(t) \leq b_i t^{u_i} e^{w_i t} \quad \text{for } t > s_0 \text{ and } i = 1, 2.$$

Then the sets

$$X_t = \{(x_1, x_2) : d_1(x_1)^p + d_2(x_2)^p < t^p\}$$

are well balanced.

Proof. For $s \in [0, 1]$, we have

$$X_t \supset X(1)_{(1-s^p)^{1/p} t} \times X(2)_{st}$$

and

$$\text{vol}(X_t) \geq v_1((1 - s^p)^{1/p} t) v_2(st) \gg (1 - s^p)^{u_1/p} s^{u_2} t^{u_1 + u_2} \exp(\kappa(s)t),$$

where $\kappa(s) = w_1(1 - s^p)^{1/p} + w_2 s$. It follows from the convexity of κ that for some $s_0 \in (0, 1), \kappa(s_0) > w_1, w_2$. This implies the claim. ☐

Typically, the averages defined by the distances $\sum_i a_i d_i$ are not balanced. However, we have the following.

Proposition 7.10. *Under the assumptions of Proposition 7.9, there exist $\alpha_1, \alpha_2 > 0$ such that the sets*

$$X_t = \{(x_1, x_2) : \alpha_1 d_1(x_1) + \alpha_2 d_2(x_2) < t\}$$

are balanced.

Proof. Rescaling the distance functions d_i by positive constants, we may assume that $\alpha_1 = \alpha_2 = 1$ and that the volume growth rates are equal, namely, $w_1 = w_2$. We have $X_t \supset X(1)_{t/2} \times X(2)_{t/2}$, and

$$\operatorname{vol}(X_t) \geq v_1(t/2) v_2(t/2) \gg t^{u_1 + u_2} \exp(w_1 t).$$

This implies the claim unless $u_1 = u_2 = 0$. In this case,

$$\operatorname{vol}(X_t) \gg \int_{t - d_2(x_2) > s_0} e^{w_1(t - d_2(x_2))} \, dx_2.$$

Since $v_2(t) \gg e^{w_2 t}$, we have $\int_{X(2)} e^{-w_2 d_2(x_2)} dx_2 = \infty$. This implies the proposition.   □

7.3 ADMISSIBLE, WELL-BALANCED, AND BOUNDARY-REGULAR FAMILIES

The present subsection is devoted to the proof of Theorem 3.18. Before recalling its formulation, we observe that when G_t is an admissible family, Haar measure m_G can be disintegrated as $m_G = m_G|_{G_{t_0}} + \int_{t_0}^{\infty} m_t \, dt$, where m_t is a measure supported on $\partial G_t, t > t_0$. Indeed, the disintegration formula holds by Proposition 3.13. We can now state the following.

Theorem 3.18. *Let $G = G(1) \cdots G(N)$ be an S-algebraic group and let ℓ_i denote the standard $CAT(0)$-metric on either the symmetric space $X(i)$ or the Bruhat-Tits building $X(i)$ associated to $G(i)$. For $1 < p < \infty$ and $u_i \in X(i)$, define*

$$G_t = \{(g_1, \ldots, g_N) : \sum_i \ell_i(u_i, g_i u_i)^p < t^p\}.$$

1. *There exist $\alpha, \beta > 0$ such that for every nontrivial projection $G \to L$ onto a factor group,*

$$m_G(G_t \cap \operatorname{proj}^{-1}(L_{\alpha t})) \ll e^{-\beta t} \cdot m(G_t),$$

   namely, the averages β_t are well balanced $(t \geq t_0)$.

2. *If G has at least one Archimedean factor and $p \leq 2$, then the family G_t is admissible and there exist $\alpha, \beta > 0$ such that for every nontrivial projection $G \to L$,*

$$m_t(\partial G_t \cap \operatorname{proj}^{-1}(L_{\alpha t})) \ll e^{-\beta t} \cdot m_t(\partial G_t),$$

   namely, the averages β_t are boundary-regular $(t > t_0)$.

In the proof, we use the following estimate.

Lemma 7.11. *With notation as in Theorem 3.18, there exist $r \geq 0$ and $\eta > 0$ such that for every $t \gg 0$,*

$$t^r e^{\eta t} \ll m_G(G_t) \ll t^r e^{\eta t}.$$

Proof. Note that the stabilizer of u_i is a maximal compact subgroup K_i of $G(i)$. For Archimedean factors, one can choose a Cartan subgroup A_i of $G(i)$ equipped with a scalar product such that the map $a \mapsto a \cdot u_i, a \in A_i$, is an isometry. Then the Cartan decomposition $G(i) = K_i A_i^+ K_i$ holds, and a Haar measure on $G(i)$ is given by $dk_1 dv_i(a) dk_2$ where

$$dv_i(a) = \prod_{\alpha \in \Sigma_i^+} \sinh(\alpha(a))^{n_{i,\alpha}} da, \tag{7.6}$$

Σ_i^+ denotes the set of positive roots, and $n_{i,\alpha}$ is the dimension of the root space. For non-Archimedean factors, there exists a lattice A_i in the centralizer of a maximal split torus, equipped with a scalar product such that the map $a \mapsto a \cdot u_i, a \in A_i$, is an isometry and $G(i) = K_i A_i^+ K_i$. Let

$$dv_i = \sum_{a \in A_i^+} \mathrm{vol}(K_i a K_i) \delta_a.$$

Note that

$$q_i^{2\rho_i(a)} \ll \mathrm{vol}(K_i a K_i) \ll q_i^{2\rho_i(a)} \tag{7.7}$$

where q_i is the order of the residue field and $2\rho_i$ is the sum of positive roots. Consider a measure $v = \otimes_i v_i$ on $A^+ = \prod_i A_i^+$, and set $d(a) = (\sum_i \|a_i\|_i^p)^{1/p}$. Let $\tilde{A}^+ = \prod_i (A_i^+ \otimes \mathbb{R}), d\tilde{v}_i(a) = q_i^{2\rho_i(a)} da$ for non-Archimedean factors, $\tilde{v}_i = v_i$ for Archimedean factors, and $\tilde{v} = \otimes_i \tilde{v}_i$. It follows from (7.7) that there exists $c > 0$ such that for $t \gg 0$,

$$\tilde{v}(\{a \in \tilde{A} : d(a) < t - c\}) \ll v(\{a \in A : d(a) < t\})$$

$$\ll \tilde{v}(\{a \in \tilde{A} : d(a) < t + c\}).$$

Hence, the proof is reduced to estimation of the integral

$$I(t) \stackrel{\mathrm{def}}{=} \int_{a \in \tilde{A}^+ : d(a) < t} d\tilde{v}(a).$$

We decompose the measure $\tilde{v}$ as a linear combination of $e^{\chi(a)} da$ for some characters χ of $\tilde{A}$. In particular, one of these characters is given by $2\rho = \sum_i (\log q_i) 2\rho_i$, where we set $\log q_i = 1$ for Archimedean factors. Let

$$\eta = \sup\{2\rho(a) : a \in \tilde{A}^+, d(a) < 1\}.$$

It follows from the proprieties of the Laplace transform that for some $c > 0$ and $r \geq 0$,

$$\int_{a \in \tilde{A}^+ : d(a) < t} e^{2\rho(a)} da \sim ct^r e^{\eta t} \quad \text{as } t \to \infty. \tag{7.8}$$

On the other hand, since the supremum η is achieved in the interior of $\tilde{A}^+$, we have

$$\sup\{\chi(a) : a \in \tilde{A}^+, d(a) < 1\} < \eta$$

for all the other characters χ appearing in the decomposition of $\tilde{v}$. Then

$$\int_{a \in \tilde{A}^+ : d(a) < t} e^{\chi(a)} da = O\left(e^{(\eta - \epsilon)t}\right)$$

for some $\epsilon > 0$, and we deduce from (7.8) that

$$I(t) \sim ct^r e^{\eta t} \quad \text{as } t \to \infty.$$

This implies the claim. $\square$

Proof of Theorem 3.18. To prove part 1, let $G = LL'$ be a nontrivial decomposition of G. Using Lemma 7.11, we deduce that for $s, \epsilon \in (0, 1)$ and $t \gg 0$, we have

$$G_t \supset L_{(1-s^p)^{1/p}t} \times L'_{st},$$

and

$$m_G(G_t) \geq m_L(L_{(1-s^p)^{1/p}t}) m_{L'}(L'_{st}) \gg_\epsilon \exp(\kappa(s, \epsilon)t),$$

where $\kappa(s, \epsilon) = (\eta - \epsilon)(1 - s^p)^{1/p} + (\eta' - \epsilon)s$. It follows from the convexity of $\kappa(\cdot, \epsilon)$ that for some $s_0, \epsilon_0 \in (0, 1)$, we have $\kappa(s_0, \epsilon_0) > \eta, \eta'$. Hence, there exists $\beta > 0$ such that for every $t \gg 0$,

$$m_L(L_t) \ll e^{-\beta t} \cdot m_G(G_t). \tag{7.9}$$

For $\alpha > 0$ and $t \gg 0$, we have

$$m_G(G_t \cap L_{\alpha t} L') \leq m_L(L_{\alpha t}) m_{L'}(L'_t) \ll m_L(L_{\alpha t}) e^{-\beta t} \cdot m_G(G_t).$$

This implies that the averages in question are well balanced.

As to part 2, namely, admissibility, the property that

$$\mathcal{O}_\epsilon G_t \mathcal{O}_\epsilon \subset G_{t+c\epsilon}, \quad \text{for some } c > 0 \text{ and every } \epsilon, t > 0,$$

follows from the triangle inequalities for d_i and the L^p-norm. Now we show that

$$m_G(G_{t+\epsilon}) \leq (1 + c\epsilon) m_G(G_t) \quad \text{for some } c > 0 \text{ and every } t > 0, \epsilon \in (0, 1). \tag{7.10}$$

Let $d(g) = (\sum_i \ell_i (u_i, gu_i)^p)^{1/p}$. Write $G = G^a G^f$, where G^a is an Archimedean factor and G^f is its complement. We denote by m_{G^a} and m_{G^f} the corresponding Haar measures. Setting

$$v(t) = m_{G^a}(G_t^a) \quad \text{and} \quad w(t, g) = v((t^p - d(g)^p)^{1/p}),$$

we have

$$m_G(G_t) = \int_{g \in G^f : d(g) < t} w(t, g) \, dm_{G^f}(g).$$

It will be convenient to set $w(t, g) = 0$ for $d(g) > t$. We claim that the function v is differentiable and

$$v'(t) \ll \max\{t, v(t)\} \quad \text{for all } t > 0. \tag{7.11}$$

To prove this, we consider the Cartan decomposition $G^a = K A^+ K$ (as in the proof of Lemma 7.11) and introduce polar coordinates $(r, \omega) \in \mathbb{R}^+ \times S^+$ on the Lie algebra of A. The Haar measure is given by $\xi(r, \omega) \, dk_1 dr d\omega dk_2$ with explicit density ξ (see (7.6)). We have

$$m_{G^a}(G^a_{t+\epsilon}) - m_{G^a}(G^a_t) = \int_{S^+} \int_t^{t+\epsilon} \xi(r, \omega) \, dr d\omega = \epsilon \int_{S^+} \xi(\sigma(\omega), \omega) \, d\omega$$

(7.12)

for some $\sigma(\omega) \in [t, t + \epsilon]$. Since $\xi(r, \omega) \ll r$ for $r \in [0, 1]$, this implies (7.11) for $t \in [0, 1]$. To establish (7.11) for $t > 1$, we use the property of the function ξ stated in (7.1).

We fix a parameter $\kappa \in (0, 1)$. It follows from (7.11) that for $d(g) < \kappa t$ and $t \gg 0$,

$$w'(t, g) = v'((t^p - d(g)^p)^{1/p}) t^{p-1} (t^p - d(g)^p)^{(1-p)/p}$$
$$\leq v((t^p - d(g)^p)^{1/p}) = w(t, g).$$

It also follows from (7.11) and Lemma 7.11 that for some $\eta > 0$ and $t > 0$, $v'(t) \ll t e^{\eta t}$ and hence for $d(g) \geq \kappa t$,

$$w'(t, g) = t^{p-1} (t^p - d(g)^p)^{(2-p)/p} \exp(\eta(t^p - d(g)^p)^{1/p})$$
$$\leq t \exp(\eta(1 - \kappa^p)^{1/p} t) \overset{\text{def}}{=} \phi(t).$$

By the mean value theorem,

$$m_G(G_{t+\epsilon}) - m_G(G_t) = \epsilon \int_{G^f} w'(\theta, g) dm_{G^f}(g),$$

where $\theta = \theta(t, \epsilon, g) \in [t, t + \epsilon]$. Applying the above estimate, we have

$$\int_{G^f} w'(\theta, g) dm_{G^f}(g) = \int_{g \in G^f : d(g) < t + \epsilon} \max\{\phi(\theta), w(\theta, g)\} dm_{G^f}(g)$$

$$\leq m_{G^f}(G^f \cap G_{t+\epsilon}) \phi(t + \epsilon) + \int_{G^f} w(t + \epsilon, g) dm_{G^f}(g).$$

Taking κ sufficiently close to 1, we arrange using (7.9) that the last term is $O(m_G(G_{t+\epsilon}))$. This implies that

$$\int_{G^f} w'(\theta, g) dm_{G^f}(g) = O(m_G(G_{t+\epsilon})),$$

and we conclude that the family G_t is admissible.

We have the following decomposition of Haar measure on G^a: $m_{G^a} = \int_0^\infty m_{a,t} dt$, where $m_{a,t}$ is a measure which is supported on ∂G^a_t. Note that for $t > 0$,

$$m_{a,t}(\partial G^a_t) = v'(t).$$

The Haar measure on G has the decomposition $m_G = \int_0^\infty m_t \, dt$, where

$$dm_t(h, g) = t^{p-1} (t^p - d(g)^p)^{1/p-1} dm_{a,(t^p-d(g)^p)^{1/p}}(h) dm_{G^f}(g)$$

for $(h, g) \in G^a \times G^f$. Hence, we have

$$m_t(\partial G_t) = \int_{g \in G^f : d(g) < t} w'(t, g) dm_{G^f}(g).$$

and by Lemma 7.11,

$$m_{G^a}(G_t^a) = \int_{S^+} \int_0^t \xi(s,\omega) \, ds \, d\omega \ll t^r e^{\eta t},$$

where $\eta = \max\{2\rho(\omega) : \omega \in S^+\}$ and 2ρ is the sum of positive roots of A. Choosing a small neighborhood U of ω_0 satisfying $2\rho(\omega_0) = \eta$, we deduce using (7.12) that for every $\delta > 0$ and $t \gg 0$,

$$m_{G^a}(G_{t+\epsilon}^a) - m_{G^a}(G_t^a) \geq \epsilon \int_U \xi(\sigma(\omega),\omega) \, d\omega \gg \epsilon \cdot e^{(\eta-\delta)t}.$$

This implies that for every $\delta > 0$ and $t \gg 0$, $v'(t) \gg v((1-\delta)t)$. Hence, for $\alpha \in (0,1)$ and $t \gg 0$,

$$m_t(\partial G_t) \geq \int_{g \in G^f : d(g) < \alpha t} w'(t,g) \, dm_{G^f}(g) \qquad (7.13)$$

$$\gg \int_{g \in G^f : d(g) < \alpha t} w(t,g) \, dm_{G^f}(g) \geq m_G(G_{\alpha t}).$$

For $\alpha \in (0,1)$ and $t \gg 0$,

$$m_t(\partial G_t \cap G_{\alpha t}^a G^f) = \int_{g \in G^f : (1-\alpha^p)^{-1/p} t < d(g) < t} w'(t,g) \, dm_{G^f}(g)$$

$$\ll \int_{g \in G^f : (1-\alpha^p)^{-1/p} t < d(g) < t} \max\{\varphi(t), w(t,g)\} \, dm_{G^f}(g)$$

$$\leq m_{G^f}(G^f \cap G_t)\varphi(t) + m_G(G_t \cap G_{\alpha t}^a G^f).$$

Also, for every nontrivial simple factor $G^f \to L$,

$$m_t(\partial G_t \cap G^a \mathrm{proj}^{-1}(L_{\alpha t})) = \int_{g \in \mathrm{proj}^{-1}(L_{\alpha t}), d(g) < t} w'(t,g) \, dm_{G^f}(g)$$

$$\ll \int_{g \in \mathrm{proj}^{-1}(L_{\alpha t}), d(g) < t} \max\{\varphi(t), w(t,g)\} \, dm_{G^f}(g)$$

$$\leq m_{G^f}(\mathrm{proj}^{-1}(L_{\alpha t}) \cap G_t) \cdot \varphi(t)$$

$$+ m_G(G_t \cap G^a \mathrm{proj}^{-1}(L_{\alpha t})).$$

Now boundary regularity follows from (7.9), part 1 of the theorem, and (7.13). $\square$

7.4 ADMISSIBLE SETS ON PRINCIPAL HOMOGENEOUS SPACES

We now consider sets defined by a distance on principal homogeneous spaces, a case of great intrinsic interest which has already appeared naturally among the examples discussed in Chapter 2.

Proposition 7.12. *Let G be a connected semisimple Lie group with finite center, $\rho : G \to GL(V)$ an irreducible representation, and $v_0 \in V$ with a compact stabilizer. We fix a norm on V and set*

$$G_t = \{g \in G : \log \|\rho(g)v_0\| < t\}.$$

Let p denote the projection on the highest weight space. If $0 \notin p(\rho(K)v_0)$ for a maximal compact subgroup K of G, then the sets G_t are admissible.

In particular, the proposition applies to the following examples discussed in §2.3. The group $G = \mathrm{SL}_k(\mathbb{R})$ acting on the space $W_{n,k}$ of homogeneous polynomials of degree n and $f_0 \in W_{n,k}$ is such that $f(x) \neq 0$ for all $x \in \mathbb{R}^k \setminus \{0\}$.

In the proof we use the following lemma.

Lemma 7.13. (cf. [EMS, Lem. A.4]). *Let $\lambda_i(\omega) \in \mathbb{R}$ and $v_i(k) \in V$ depend on parameters ω, k, and for every $s \geq 0$,*

$$\exp(s \max_i \lambda_i(\omega)) \ll \left\| \sum_i e^{s\lambda_i(\omega)} v_i(k) \right\| \ll \exp(s \max_i \lambda_i(\omega))$$

uniformly on ω, k. Then there exists $T_0 > 0$ such that for every $T > T_0$, the set

$$\left\{ s \geq 0 : \left\| \sum_i e^{s\lambda_i(\omega)} v_i(k) \right\| < T \right\}$$

is an interval $[0, r(T, \omega, k)]$ and

$$r((1 + \epsilon)T, \omega, k) - r(T, \omega, k) \ll \epsilon \tag{7.14}$$

uniformly on $\omega, k, \epsilon \in (0, 1)$, and $T > T_0$.

We note that the statement of Lemma A.4 in [EMS] is somewhat weaker than the statement above, but the proof there implies the lemma in this generality.

Proof of Proposition 7.12. We fix a Cartan decomposition $G = KA^+K$. For a weight λ of $\mathrm{Lie}(A)$, we denote by p_λ the projection on the weight space of λ. Then for $g \in k_1 \exp(a)k_2 \in G$,

$$g v_0 = \sum_\lambda e^{\lambda(a)} k_1 p_\lambda(k_2 v_0).$$

This implies that

$$\max_{\lambda, k_2} e^{\lambda(a)} \| p_\lambda(k_2 v_0) \| \ll \| g v_0 \| \ll \max_{\lambda, k_2} e^{\lambda(a)} \| p_\lambda(k_2 v_0) \|,$$

and it follows from the assumption $0 \notin p(K v_0)$ that

$$e^{\lambda_{\max}(a)} \ll \| g v_0 \| \ll e^{\lambda_{\max}(a)}. \tag{7.15}$$

It is straightforward to check that $\mathcal{O}_\epsilon G_t \subset G_{t+c\epsilon}$ for some $c > 0$. Now we show that $G_t \mathcal{O}_\epsilon \subset G_{t+c\epsilon}$ as well. For $g = k_1 \exp(a)k_2$ and h in G, we have

$$\| g h v_0 - g v_0 \| \ll \sum_\lambda e^{\lambda(a)} \| p_\lambda(k_2(h v_0 - v_0)) \| \ll e^{\lambda_{\max}(a)} \| h v_0 - v_0 \|$$

$$\ll \| g v_0 \| d(h, e).$$

This estimate implies the claim.

It remains to show that for sufficiently small $\epsilon > 0$ and for sufficiently large t,

$$m_G(G_{t+\epsilon}) - m_G(G_t) \ll \epsilon \, m_G(G_t). \tag{7.16}$$

Since (7.15) holds, we may apply Lemma 7.13 with vectors v_i given by $k_1 p_\lambda(k_2 v_0)$ with $k_1, k_2 \in K$. This implies the Lipschitz estimate (7.14), and (7.16) is deduced as in [EMS, Prop. A.5]. $\qquad\square$

Proposition 7.14. *Under the assumptions of Proposition 7.12, there exist $c > 0$, $b = 0, \ldots , \mathrm{rank}(G) - 1$, and $\alpha \in \mathbb{R}_+$ such that*

$$\mathrm{vol}(G_t) \sim c\, t^{b-1} e^{\alpha t} \quad as\ t \to \infty.$$

Proof. Fix a Cartan decomposition $G = K A^+ K$. Then the Haar measure on G is given by $\xi(a)dk_1 da dk_2$. For $k_1, k_2 \in K$, set

$$A_t(k_1, k_2) = \{a \in A^+ : \log \|k_1 a k_2\| < t\}.$$

We have

$$\mathrm{vol}(G_t) = \int_{K \times K} \int_{A_t(k_1, k_2)} \xi(a) da dk_1 dk_2.$$

By [GW, §7],

$$\int_{A_t(k_1,k_2)} \xi(a)da \sim c(k_1, k_2) t^{b(k_1, k_2)} e^{\alpha(k_1, k_2)t} \quad as\ t \to \infty, \qquad (7.17)$$

where $c(k_1, k_2) > 0$, $b(k_1, k_2) \in \mathbb{Z}_+$, and $\alpha \in \mathbb{R}_+$. Also, it follows from (7.15) that there exists $c > 0$ such that for all $k_1, k_2 \in K$ and $t \gg 0$,

$$A_{t-c}(e, e) \subset A_t(k_1, k_2) \subset A_{t+c}(e, e).$$

This implies that the exponents in (7.17) are constant and

$$t^{b-1} e^{\alpha t} \ll \int_{A_t(k_1,k_2)} \xi(a)da \ll t^{b-1} e^{\alpha t}$$

for sufficiently large t and $k_1, k_2 \in K$. Now the claim follows from the dominated convergence theorem. $\square$

7.5 TAUBERIAN ARGUMENTS AND HÖLDER CONTINUITY

Finally, we will now establish the Hölder-admissibility property of averages defined by a regular proper function on algebraic varieties with a regular volume form. Our approach uses the following Tauberian theorem which is proved using the argument in [CT, Thm. A.1].

Proposition 7.15. *Let $v : [0, \infty) \to [0, \infty)$ and $f(s) = \int_0^\infty x^{-s} v(x)\, dx$.*

1. *Let $v(t)$ be increasing for sufficiently large t. Assume that the integral $f(s)$ converges for $\mathrm{Re}(s) \gg 0$, admits meromorphic continuation to $\mathrm{Re}(s) > a - \delta_0$, and in this domain has a unique pole $s = a$ of multiplicity b and satisfies*

$$\left| f(s) \frac{(s - a)^b}{s^b} \right| = O((1 + \mathrm{Im}(s))^\kappa)$$

for some $\kappa > 0$. Then

$$v(t) = t^{a-1} P(\log t) + O_\delta(t^{a-1-\delta}) \quad as\ t \to \infty$$

for some nonzero polynomial P and $\delta > 0$.

2. *Let $v(t)$ be increasing for sufficiently small t. Assume that the integral $f(s)$ converges for $\mathrm{Re}(s) \ll 0$, admits meromorphic continuation to $\mathrm{Re}(s) < a + \delta_0$, and in this domain has a unique pole $s = a$ of multiplicity b and satisfies*

$$\left| f(s)\frac{(s-a)^b}{s^b} \right| = O((1 + \mathrm{Im}(s))^\kappa)$$

for some $\kappa > 0$. Then

$$v(t) = t^{a-1} P(\log t) + O_\delta(t^{a-1+\delta}) \quad \text{as } t \to 0^+$$

for some nonzero polynomial P and $\delta > 0$.

Proof. The first statement is essentially proved in [CT] (in the context of a Dirichlet series), and the second statement is proved similarly. We give a sketch of the proof for the second statement.

For negative $a' < a$, define

$$w_k(t) = \frac{(-1)^{k+1}k!}{2\pi i} \int_{a'+i\mathbb{R}} f(s)t^s \frac{ds}{s^{k+1}}.$$

This integral is absolutely convergent for $k > \kappa$. Applying the Cauchy formula for the region $a' < \mathrm{Re}(s) < a + \delta$, $\mathrm{Im}(s) \le S$ with $S \to \infty$, we deduce that for some nonzero polynomial P_k,

$$w_k(t) = t^a P_k(\log t) + O(t^{a+\delta}) \quad \text{as } t \to 0^+. \tag{7.18}$$

It follows from the formula

$$\int_{a'+i\mathbb{R}} \lambda^s \frac{ds}{s^{k+1}} = \begin{cases} -\frac{2\pi i}{k!}(\log \lambda)^k & 0 < \lambda \le 1, \\ 0 & \lambda > 1 \end{cases}$$

that

$$w_k(t) = (-1)^k \int_t^\infty (\log(t/x))^k v(x)dx = \int_t^\infty (\log(x/t))^k v(x)dx.$$

Now we derive an asymptotic expansion for w_{k-1} assuming that (7.18) holds. By the intermediate value theorem, for every $t > 0$ and $\eta \in (0, 1)$,

$$w_{k-1}(t) \le \frac{\int_t^\infty \left((\log(x/t(1-\eta)))^k - (\log(x/t))^k \right) v(x)dx}{-k\log(1-\eta)}$$

$$\le \frac{w_k(t(1-\eta)) - w_k(t)}{-k\log(1-\eta)}$$

and

$$w_{k-1}(t) \ge \frac{\int_{t(1+\eta)}^\infty \left((\log(x/t))^k - (\log(x/t(1+\eta)))^k \right) v(x)dx}{k\log(1+\eta)}$$

$$+ \frac{1}{\log(1+\eta)} \int_t^{t(1+\eta)} (\log(x/t))^k v(x)dx$$

$$\ge \frac{w_k(t) - w_k(t(1+\eta))}{k\log(1+\eta)}.$$

Taking $\eta = t^\epsilon$ with small $\epsilon > 0$ and using (7.18), we deduce from the above estimates that

$$w_{k-1}(t) = t^a P_{k-1}(\log t) + O\left((\log t)^{\deg P_k} t^{a+\epsilon} + t^{a+\delta-\epsilon}\right) \quad \text{as } t \to 0^+$$

for some nonzero polynomial P_{k-1}. Now using an inductive argument, we deduce that (7.18) holds for all $k \geq 0$. To complete the proof, we observe that for small $t > 0$ and $\eta \in (0, 1)$,

$$\frac{1}{t\eta}(w_0(t(1-\eta)) - w_0(t)) \leq v(t) \leq \frac{1}{t\eta}(w_0(t) - w_0(t(1+\eta))).$$

Setting $\eta = t^\epsilon$ with small $\epsilon > 0$, we deduce the required asymptotic expansion for $v(t)$ from the asymptotic expansion for $w_0(t)$. $\qquad\square$

We will now employ Proposition 7.15 and prove the Hölder continuity of the volume function in the context of algebraic functions on algebraic varieties.

Theorem 7.16. *Let X be a real algebraic affine variety equipped with a regular volume form ω and $\Psi : X \to \mathbb{R}$ a regular proper function. Then the function $g(t) = \int_{\Psi(x)<t} d\omega$ is uniformly Hölder-continuous on finite intervals.*

Proof. We equip X with the Riemannian metric coming from the ambient Euclidean space.

At regular values t of Ψ,

$$g'(t) = \int_{\Psi^{-1}(t)} \frac{dv_t}{\|\nabla\Psi\|},$$

where v_t is the induced measure on the fiber $\Psi^{-1}(t)$. We claim that for t in a neighborhood of an isolated critical value t_0,

$$g'(t) \ll |t - t_0|^{-r} \tag{7.19}$$

for some $r > 0$. Let Z be the set of critical points of Ψ in $\Psi^{-1}(t_0)$. By Lojasiewicz's inequality (see, e.g., [BM, Thm. 6.4]),

$$\|(\nabla\Psi)_x\| \gg d(x, Z)^r$$

for some positive r. For $x \in \Psi^{-1}(t)$ and $z \in \Psi^{-1}(t_0)$,

$$d(x, z) \gg |\Psi(x) - \Psi(z)| = |t - t_0|.$$

This implies (7.19), and by the intermediate value theorem, for $t_0 < t_1 < t_2$ in a neighborhood of t_0,

$$|g(t_2) - g(t_1)| \ll |t_1 - t_0|^{-r} \cdot |t_2 - t_1|. \tag{7.20}$$

Next, we show that g is Hölder-continuous at critical values of Ψ. For $c \in \mathbb{R}$, we consider the function v defined by $v(t) = \int_{c \leq \Psi < c+t} d\omega$ for $t < 1$ and $v(t) = 0$ for $t \geq 1$, and its transform

$$f(s) = \int_0^\infty t^{-s} v(t) dt = \int_{c \leq \Psi(x) < c+1} \frac{1 - (\Psi(x) - c)^{1-s}}{1 - s} d\omega(x),$$

which is absolutely convergent for $\text{Re}(s) < 1$. Applying Hironaka resolution of singularities to the function $F(x) = \Psi(x) - c$, we deduce that there exists an atlas of maps $\phi_i : (-1, 1)^d \to U_i$, U_i is open in X, such that ϕ_i are diffeomorphisms on sets of full measure, and

$$F(\phi_i(x)) = x^{\alpha_i} F_i(x) \quad \text{and} \quad d\omega(\phi_i(x)) = x^{\beta_i} \rho_i(x) dx,$$

where x^{α_i} and x^{β_i} denote monomials and F_i and ρ_i are positive smooth functions. Let $\{\eta_i\}$ be a partition of unity subordinate to the cover $\{U_i\}$ such that

$$\sum_i \eta_i = 1 \text{ on } \{c \le \Psi(x) \le c + 1/2\} \text{ and } \text{supp}(\eta_i) \subset \{\Psi(x) < c + 1\}.$$

We have

$$\int_{c \le \Psi(x) < c+1} (\Psi(x) - c)^{1-s} d\omega(x) \tag{7.21}$$

$$= \sum_i \int_{x \in (-1,1)^d : x^{\alpha_i} \ge 0} x^{\beta_i + (1-s)\alpha_i} F_i(x)^{1-s} \rho_i(x) \eta_i(\phi_i(x)) dx + \xi(s),$$

where $\xi(s)$ is an integral over the region $\Psi(x) > c + 1/2$, hence, holomorphic, and the other integrals can be meromorphically continued using integration by parts:

$$\int_{(0,1)^d} x^{\beta_i + (1-s)\alpha_i} F_i(x)^{1-s} \rho_i(x) \eta_i(\phi_i(x)) dx$$

$$= \left(\prod_j (1 + \beta_i + (1-s)\alpha_{i,j}) \right)^{-1}$$

$$\times \int_{(0,1)^d} x^{1+\beta_i + (1-s)\alpha_i} (F_i(x)^{1-s} \rho_i(x) \eta_i(\phi_i(x)))' dx.$$

Therefore (7.21) implies that the conditions of Proposition 7.15(2) are satisfied, and hence,

$$v(t) = t^{a-1} P(\log t) + O(t^{a-1+\delta}) \quad \text{as } t \to 0^+$$

for some $a \ge 1$. If $a = 1$, then $\text{vol}(\{x : \Psi(x) = c\}) > 0$, but the set $\{x : \Psi(x) = c\}$ is a proper algebraic subvariety of X. Hence, $a > 1$, and we deduce the Hölder estimate

$$\text{vol}(\{x : c \le \Psi(x) < c+t\}) \ll t^\alpha, \tag{7.22}$$

with $\alpha < a - 1$.

Since g is a polynomial function, it has only finitely many critical values. The function g is C^1 on the set of regular values. Hence, it remains to show that g is Hölder-continuous in a neighborhood of a critical values. For instance, consider the case when t_1 and t_2 are in the neighborhood of a critical value t_0 and $t_0 < t_1 < t_2$. The other cases are treated similarly. We have

$$|g(t_2) - g(t_1)| \ll (t_1 - t_0)^\alpha + (t_2 - t_0)^\alpha \ll (t_1 - t_0)^\alpha + (t_2 - t_1)^\alpha. \tag{7.23}$$

When $(t_1 - t_0)^{-r} \le (t_2 - t_1)^{-1/2}$, the Hölder estimate follows from (7.20), and when the opposite inequality holds, the Hölder estimate follows from (7.23). This completes the proof. $\square$

Using Proposition 7.15, we can now obtain the following volume asymptotics.

Theorem 7.17. *Let X be a real algebraic affine variety equipped with a regular volume form ω, $\Psi : X \to [1, \infty)$ a regular proper function, and*

$$v(t) = \mathrm{vol}(\{x \in X : \ \Psi(x) < t\}).$$

Then for some $a \geq 1$, a nonzero polynomial P, and $\delta \in (0, 1)$, we have

$$v(t) = t^{a-1} P(\log t) + O_\delta(t^{a-1-\delta}) \quad \text{as } t \to \infty.$$

Proof. First note that since the function $\theta(t) = \max\{\|x\| : \Psi(x) < t\}$ is semialgebraic, there exists $M > 0$ such that $\theta(t) \ll t^M$ for sufficiently large t. This implies that for some $N > 0$ and $t \gg 0$, we have $v(t) \ll t^N$.

Now let $w(t) = v(t) - v(1)$ for $t \geq 1$ and $w(t) = 0$ for $t < 1$. Consider the transform of w:

$$f(s) = \int_0^\infty t^{-s} w(t) dt,$$

which is convergent for $\mathrm{Re}(s) > N + 1$, and

$$f(s) = (s - 1)^{-1} \int_X \Psi(x)^{-s+1} d\omega(x).$$

Applying Hironaka resolution of singularities, we may assume X is a semialgebraic subset of a smooth projective variety Y, and there exists an atlas of maps $\phi_i : (-1, 1)^d \to Y$, U_i is open in Y, such that ϕ_i are diffeomorphisms on sets of full measure, and $\phi_i^{-1}(U_i \cap X)$ is a union of quadrants, and

$$\Psi(\phi_i(x))^{-1} = x^{\alpha_i} \Psi_i(x), \qquad d\omega(\phi_i(x)) = x^{\beta_i} \rho_i(x) dx,$$

where x^{α_i} and x^{β_i} denote monomials and F_i and ρ_i are smooth functions nonvanishing on $(-1, 1)^d$. Let $\{\eta_i\}$ be a partition of unity subordinate to the cover $\{U_i\}$. Then

$$\int_X \Psi(x)^{-s+1} d\omega(x) = \sum_i \int_{\phi_i^{-1}(U_i \cap X)} x^{(s-1)\alpha_i + \beta_i} \Psi_i(x)^{s-1} \rho_i(x) \eta_i(\phi_i(x)) dx.$$

Integrating by parts, we deduce that this expresion has meromorphic continuation and satisfies the conditions of Proposition 7.15. This implies the claim. $\qquad\square$

Theorem 7.18. *Let X be a real algebraic affine variety equipped with a regular volume form ω and $\Psi : X \to [1, \infty)$ a regular proper function. Then for some $\beta > 0$, the function $g(t) = \int_{\Psi < t} d\omega$ satisfies*

$$g((1 + \epsilon)t) - g(t) \ll \epsilon^\beta \max\{1, g(t)\} \quad \text{for all } \epsilon \in (0, 1) \text{ and } t \geq 0.$$

Proof. On finite intervals, this has already been proved in Theorem 7.16. Since Ψ is regular, it has only finitely many critical points. Thus, it remains to consider an interval $t \gg 0$ which contains no critical points. It follows from Theorem 7.17 that for $t \geq \epsilon^{-\alpha}$ with arbitrary $\alpha > 0$, we have

$$g((1 + \epsilon)t) - g(t) \ll \epsilon^\beta g(t),$$

where $\beta > 0$ depends on α. To prove the estimate for $t < \epsilon^{-\alpha}$, we use the formula

$$g'(t) = \int_{\Psi=t} \|(\nabla\Psi)_x\|^{-1} \, d\omega_t,$$

where ω_t is the regular volume form on $\{\Psi = t\}$ induced by ω. Since the function $t \mapsto \max\{\|(\nabla\Psi)_x\|^{-1} : \Psi(x) = t\}$ is semialgebraic, it follows that there exists $M > 0$ such that for $\Psi(x) \gg 0$,

$$\|(\nabla\Psi)_x\|^{-1} \ll \Psi(x)^M.$$

Similarly, for $\Psi(x) \gg 0$,

$$\|x\| \ll \Psi(x)^M.$$

This implies that for some $N > 0$ and $t \gg 0$, we have

$$g'(t) \ll t^N.$$

Then when $0 \ll t < \epsilon^{-\alpha}$, with $\alpha < 1/N$, we have Hölder estimate

$$g(t + \epsilon) - g(t) \ll \epsilon^{1-\alpha N}.$$

Hence, the claim follows. $\square$

Finally, we combine the foregoing arguments to prove Hölder admissibility of families defined by a height function on a product of affine varieties and, in particular, on an S-algebraic group (as stated in Theorem 3.16).

Theorem 7.19. *Let $X = X(1) \cdots X(N)$ be a product of affine varieties $X(i)$ over local fields equipped with regular volume forms. We denote by $\| \cdot \|_i$ either the Euclidean norm for Archemedian places or the* max *norm for non-Archemedean places and set*

$$X_t = \{(x_1, \dots, x_N) : \sum_i \log \|x_i\|_i < t\}.$$

If at least one of the factors of X is Archimedean and noncompact, then the function $t \mapsto \mathrm{vol}(X_t)$ is uniformly locally Hölder-continuous.

Proof. Setting $v_i(t) = \mathrm{vol}(\{x_i : \log \|x_i\| < t\})$, the claim is deduced by applying Propositions 7.8 and 7.7 inductively. Using restriction of scalars, we can assume that all Archimedean factors are real. For Archimedean v_i, the assumption of Proposition 7.8 follows from Theorem 7.18 and the assumption of Proposition 7.7 follows from Proposition 7.17. Hence, it remains to verify the assumption of Proposition 7.7 at non-Archimedean places. In this case, it follows from [D] that $\int_{X(i)} \|x\|_i^s \, d\omega_i(x)$ is a rational function of q_i^s and q_i^{-s}, where q_i is the order of the residue field. Hence,

$$w_i(n) := \mathrm{vol}(\{x \in X(i) : \|x_i\|_i = q_i^n\}) = \sum_j p_{ij}(n)q_i^{a_{ij}n}$$

for rational polynomials p_{ij} and $a_{ij} \in \mathbb{Z}$. This implies that the function $v_i(t) = \sum_{n<t} w_i(n)$ satisfies $v_i(t + 1) \ll v_i(t)$ for sufficiently large t.

Now the claim follows from Propositions 7.8 and 7.7. $\square$

Chapter Eight

Comments and complements

In the present chapter we give a formula for the error term in the lattice point problem, present an example of an action where exponentially fast almost sure convergence holds but equidistribution fails, and comment on the existence of balanced averages.

8.1 LATTICE POINT–COUNTING WITH EXPLICIT ERROR TERM

Let us state the following explicit error estimate in the lattice point–counting problem in admissible domains, which follows from the results above. It is convenient to denote by Σ_{p^+} the subset of the positive-definite spherical functions on an S-algebraic group G that are in $L^{p+\varepsilon}(G)$ for every $\varepsilon > 0$. If

$$\int_G |\phi|^{p+\varepsilon} \, dm_G(g) \leq C(\varepsilon) < \infty$$

is independent of $\phi \in \Sigma_{p^+}$, we say that the L^{p^+}-spherical spectrum is uniformly bounded.

Corollary 8.1. Quantitative estimate in the lattice point–counting problem.
Let G be an S-algebraic group as in Definition 3.4. Let Γ be a lattice subgroup and G_t an admissible family, both contained in G^+. If G_t are well balanced or the lattice is irreducible, then

1. The number of lattice points in $\Gamma_t = G_t \cap \Gamma$ is estimated by (for all $\varepsilon > 0$)

$$\left| \frac{|\Gamma_t|}{m_G(G_t)} - 1 \right| \leq B_\varepsilon \exp\left(-\frac{(\theta - \varepsilon)t}{\varrho_0 + 1} \right),$$

where θ is given by

$$\theta = \liminf_{t \to \infty} -\frac{1}{t} \log \left\| \pi_{G/\Gamma}(\beta_t) \right\|_{L^2_0(G/\Gamma)},$$

and ϱ_0 is the real dimension of the Archimedean component.

2. If $\pi = \pi^0_{G/\Gamma}$ has a strong spectral gap so that it is an L^p-representation, then $\pi^{\otimes n_e} \subset \infty \cdot \lambda_G$ for $n_e \geq p/2$ even, and the spectral parameter θ satisfies

$$\theta \geq \frac{1}{2n_e} \liminf_{t \to \infty} \frac{1}{t} \log m_G(B_t).$$

3. *When G_t are bi-invariant under a good maximal compact subgroup of G, we can replace ϱ_0 in the estimate in part 1 by the real dimension of the symmetric space associated with the Archimedean factor. Furthermore, if the L^{p^+}-spherical spectrum is uniformly bounded, we can replace $\frac{1}{2n_e}$ by $\frac{1}{p}$ in the estimate of θ in part 2.*

We note that part 1 of Corollary 8.1 follows immediately from Lemma 6.11, taking $r = p = 2$ and ϱ_0 to be the upper local dimension. Part 2 follows immediately from the discussion in §5.2.2.2 and the fact that S-arithmetic groups as in Definition 3.4 have the Kunze-Stein property. This is well known in the real case in [C1] and was proved by A. Veca [Ve, Thm. 1] in the totally disconnected simply connected case.

We remark that for admissible families the estimate of θ in part 2 can be established using (in effect) just the radial Kunze-Stein phenomenon, which is much easier to establish than the full result. This proceeds by using the fact that admissible averages are (K, C)-radial and estimating the norm of the corresponding radialized averages directly using Proposition 5.9(2) Theorem 5.3(3) and the standard estimate of the Ξ-function.

Part 3 follows from the following two facts. First, when G_t are bi-K-invariant, the argument in Lemma 6.11 in effect takes place on $K \backslash G$, so we can consider the orbit points $K\Gamma$ in the sets $KG_t \subset K \backslash G$. In the approximation argument utilizing the local family $K\mathcal{O}_\varepsilon \subset K \backslash G$, the relevant volume estimate of $K\mathcal{O}_\varepsilon$ is now given by $\varepsilon^{\varrho_0^K}$, where ϱ_0^K is the real dimension of the symmetric space associated with the Archimedean component. Second, β_t now belong to the commutative convolution algebra of bi-K-invariant L^1-functions on G, and so $\left\| \pi_{G/\Gamma}(\beta_t) \right\|_{L_0^2(G/\Gamma)}$ can now be estimated directly using the spectral theory of this algebra, namely, the theory of positive-definite spherical functions. The estimate in question when $\pi_{G/\Gamma}^0$ is an L^{p^+}-representation is

$$\sup_{\phi \in \Sigma_{p^+}} \frac{1}{m_G(G_t)} \int_{G_t} |\phi(g)| \, dm_G(g).$$

Assuming that the L^{p^+}-spherical spectrum is uniformly bounded, we can apply Hölder's inequality to obtain the bound $C_\varepsilon m_G(G_t)^{-1/p+\varepsilon}$ for every $\varepsilon > 0$, and the desired estimate follows.

We remark that one case where one can establish that the L^{p^+}-spherical spectrum is uniformly bounded is when there exists a single positive function $\Phi_{G,p}(g) \in L^{p^+}$ which pointwise dominates all normalized positive-definite spherical functions which are in L^{p^+}. A remarkable argument due to Howe (see [HT]) establishes explicitly the existence of such a function $\Phi_{G,p}$ for (say) $G = \mathrm{SL}_n(\mathbb{R})$, $n \geq 3$, when $p = p_K(G)$ is the least p such that all nonconstant positive-definite irreducible spherical functions are in L^{p^+}. The same method of proof can be applied to all other simple algebraic groups with split rank at least two over local fields and yields a fixed positive function F_G dominating all nonconstant positive-definite spherical functions, with $F \in L^{q^+}(G)$, for some $q \geq p_K(G)$ (see [O] for the details).

Remark 8.2. Lattice point–counting problem. Corollary 8.1 constitutes a quantitative solution to the lattice point–counting problem in admissible domains in the group variety. For a systematic discussion of quantitative counting results for more general domains (including Hölder well-rounded domains and in particular sectors) and more general groups (including Adele groups), together with many applications, we refer the reader to [GN].

8.2 EXPONENTIALLY FAST CONVERGENCE
VERSUS EQUIDISTRIBUTION

In this section we give an example of a connected semisimple Lie group H without compact factors acting by translations on a homogeneous space G/Γ of finite volume, and of Haar-uniform admissible averages β_t on H such that

- equidistribution of H-orbits fails (i.e., there exist dense orbits for which the averages do not converge to the Haar measure),

- An exponential pointwise ergodic theorem holds (i.e., for almost all starting points the averages converge to the Haar measure exponentially fast).

This example was originally constructed in [GW, §12.3].
Let

$$H = H(1) \times H(2) = \mathrm{SL}_2(\mathbb{R}) \times \mathrm{SL}_2(\mathbb{R})$$

and

$$r : H \to \mathrm{SL}_{2l}(\mathbb{R})$$

be the representation of H which is a tensor product of irreducible representations of $H(1)$ and $H(2)$ of dimensions 2 and $l > 2$, respectively. We fix a norm $\|\cdot\|$ on $M_{2l}(\mathbb{R})$ and define

$$H_t = \{h \in H : \|r(h)\| < e^t\}.$$

Note that the sets H_t are not balanced, as shown in [GW].
Let $G = \mathrm{SL}_{2l}(\mathbb{R})$ and $\Gamma = \mathrm{SL}_{2l}(\mathbb{Z})$. For $x \in G/\Gamma$ and $t > 0$, consider the Radon probability measure

$$\mu_{x,t}(f) = \frac{1}{m_H(H_t)} \int_{H_t} f(h^{-1}x)dm_H(h), \quad f \in C_c(G/\Gamma).$$

Proposition 8.3.

1. *There exists $x \in G/\Gamma$ such that $\overline{Hx} = G/\Gamma$, but Haar measure $m_{G/\Gamma}$ is not an accumulation point of the family $\mu_{x,t}$, $t \to \infty$, in the weak* topology on $C_c(G/\Gamma)$.*

2. *For almost everywhere $x \in G/\Gamma$, $\mu_{x,t} \to m_{G/\Gamma}$ as $t \to \infty$ in the weak* topology. Moreover, for every $p > r \geq 1$, there exists $\theta = \theta_{p,r} > 0$ such that for $f \in L^p(m_{G/\Gamma})$ and almost everywhere $x \in G/\Gamma$,*

$$\left| \mu_{x,t}(f) - \int_{G/\Gamma} f \, dm_{G/\Gamma} \right| \leq C(x, f)e^{-\theta_{p,r}t},$$

with

$$\|C(\cdot, f)\|_{L^r(m_{G/\Gamma})} \le C\|f\|_{L^p(m_{G/\Gamma})}.$$

Proof. Part 1 was proved in [GW, §12.3].

To prove part 2, it suffices to observe that the representation of G on $L_0^2(G/\Gamma)$ has a spectral gap. Being simple, the spectral gap is strong, and so some tensor power of the representation embeds in $\infty \cdot \lambda_G$, as follows from the spectral transfer principle (see Theorem 5.3 or [N4]). Thus the same tensor power of the representation of H on $L_0^2(G/\Gamma)$ (restricted to H) embeds in $\infty \cdot \lambda_H$, and so H has a strong spectral gap as well. Therefore the desired result follows from Theorem 4.2.   □

8.3 REMARK ABOUT BALANCED SETS

The last example owes its existence to the fact that the averages considered are not balanced. Let us therefore give an easy geometric criterion for a family of sets defined by a matrix norm on a product of simple groups to be balanced.

Let $G = G(1) \cdots G(N)$ be a connected semisimple Lie group, where $G(i)$ are the (noncompact) simple factors, and let

$$\mathfrak{a} = \mathfrak{a}_1 \oplus \cdots \oplus \mathfrak{a}_N \tag{8.1}$$

be a Cartan subalgebra of G where $\mathfrak{a}_i$ are Cartan subalgebras of $G(i)$. We fix a system of simple roots $\Phi = \Phi_1 \cup \cdots \cup \Phi_N$ for $\mathfrak{a}$, where Φ_i is a system of simple roots for $\mathfrak{a}_i$, and denote by

$$\mathfrak{a}^+ = \mathfrak{a}_1^+ \oplus \cdots \oplus \mathfrak{a}_N^+$$

the corresponding positive Weyl chamber.

Let $r : G \to \mathrm{GL}_d(\mathbb{R})$ be a representaion of G. For a norm $\|\cdot\|$ on $\mathrm{M}_d(\mathbb{R})$, let

$$G_t = \{g \in G : \log \|r(g)\| < t\}.$$

Let Ψ_r be the set of weights of $\mathfrak{a}$ in the representation r, and

$$\mathfrak{p}_r = \{H \in \mathfrak{a}^+ : \lambda(H) \le 1 \quad \text{for all } \lambda \in \Psi_r\}.$$

Finally, let

$$\delta = \max\{\rho(H) : H \in \mathfrak{p}_r\},$$

where ρ denotes the half-sum of the positive roots of $\mathfrak{a}$.

We can now formulate the following.

Proposition 8.4. *The sets G_t are balanced iff the set $\{\rho = \delta\} \cap \mathfrak{p}_r$ is not contained in any proper subsum of the direct sum (8.1).*

The proposition follows from [GW, §7].

Bibliography

[A]   M. Atiyah, *Resolution of singularities and division of distributions*. Comm. Pure Appl. Math. **23** (1970), 145–150.

[Ba]   H. J. Bartels, *Nichteuklidische Gitterpunktprobleme und Gleichverteilung in linearen algebraischen Gruppen*. Comment. Math. Helv. **57** (1982), 158–172.

[Be]   M. B. Bekka, *On uniqueness of invariant means*. Proc. Amer. Math. Soc. **126** (1998), 507–514.

[BO]   Y. Benoist and H. Oh, *Effective equidistribution of S-integral points on symmetric varieties*. arXiv:0706.1621.

[BeGe] I. Bernstein and S. Gelfand, *Meromorphy of the function P^λ*. Funct. Anal. Appl. **3** (1969), 68–69.

[BM]   E. Bierstone and P. Milman, *Semianalytic and subanalytic sets*. Inst. Hautes Études Sci. Publ. Math. **67** (1988), 5–42.

[BoGa] A. Borel and H. Garland, *Laplacian and the discrete spectrum of an arithmetic group*. Amer. J. Math. **105** (1983), 309–335.

[BW]   A. Borel and N. Wallach, *Continuous Cohomology, Discrete Subgroups, and Representation of Reductive Groups*. Ann. of Math. Stud. **94**, Princeton University Press, Princeton, NJ, 1980.

[BR]   L. Bowen and C. Radin, *Optimally dense packing of hyperbolic space*. Geom. Dedicata. **104** (2004), 37–59.

[CT]   A. Chambert-Loir and Y. Tschinkel, *Fonctions zêta des hauteurs des espaces fibrés*. In: Rational points on algebraic varieties, Progr. Math. **199**, Birkhauser, Basel, 2001, pp. 71–115.

[C1]   M. Cowling, *The Kunze-Stein phenomenon*. Ann. Math. **107** (1978), 209–234.

[C2]   M. Cowling, *Sur les coefficients des représentations unitaires des groupes de Lie simples*. Analyse harmonique sur les groupes de Lie, Séminaire Nancy–Strasbourg 1975–77. Lecture Notes in Math. **739**, Springer Verlag, Berlin, 1979, pp. 132–178.

[CHH] M. Cowling, U. Haagerup and R. Howe, *Almost L^2-matrix coefficients*. J. Reine Angew. Math. **387** (1988), 97–110.

[CN] M. Cowling and A. Nevo, *Uniform estimates for spherical functions on complex semisimple Lie groups*. Geom. Funct. Anal. **11** (2001), 900–932.

[D] J. Denef, *The rationality of the Poincaré series associated to the p-adic points on a variety*. Invent. Math. **77** (1984), 1–23.

[DRS] W. Duke, Z. Rudnick, and P. Sarnak, *Density of integer points on affine homogeneous varieties*. Duke Math. J. **71** (1993), 143–179.

[EM] A. Eskin and C. McMullen, *Mixing, counting and equidistribution in Lie groups*. Duke Math. J. **71** (1993), 181–209.

[EMS] A. Eskin, S. Mozes, and N. Shah, *Unipotent flows and counting lattice points on homogeneous varieties*. Ann. Math. **143** (1997), 253–299.

[GV] R. Gangolli and V. S. Varadarajan, *Harmonic Analysis of Spherical Functions on Real Reductive Groups*. Modern Surveys in Math. **101**, Springer Verlag, New York, 1988.

[GN] A. Gorodnik and A. Nevo, *Counting lattice points*. arXiv:0903.1515.

[GW] A. Gorodnik and B. Weiss, *Distribution of lattice orbits on homogeneous varieties*. Geom. Funct. Anal. **17** (2007), 58–115.

[G] Y. Guivarc'h, *Gèneralisation d'un thèreme de von-Neumann*. C. R. Acad. Sci. Paris **268** (1969), 1020–1023.

[HC1] Harish-Chandra, *Spherical functions on a semi-simple Lie group I*. Amer. J. Math. **80** (1958), 241–310.

[HC2] Harish-Chandra, *Spherical functions on a semi-simple Lie group II*. Amer. J. Math. **80** (1958), 553–613.

[HC3] Harish-Chandra, *Harmonic analysis on reductive p-adic groups*. In: Harmonic Analysis on Homogeneous Spaces. Proc. Sympos. Pure Math. **26**, American Mathematical Society, Providence, RI, 1973, pp. 167–192.

[He1] S. Helgason, *Differential Geometry, Lie Groups and Symmetric Spaces*. Academic Press, New York, 1978.

[He2] S. Helgason, *Groups and Geometric Analysis*. Academic Press, New York, 1984.

[Ho] R. E. Howe, *On a notion of rank for unitary representations of the classical groups*. Harmonic Analysis and Group Representations, C.I.M.E., 2° ciclo, Liguori, Naples, 1982, pp. 223–232.

[HT] R. E. Howe and E. C. Tan, *Non-Abelian Harmonic Analysis*. Springer Verlag, New York, 1992.

[HM] R. E. Howe and C. C. Moore, *Asymptotic properties of unitary representa-tions*. J. Funct. Anal. **32** (1979), 72–96.

[J]   J. Jost, *Riemannian Geometry and Geometric Analysis*. Universitext. Springer Verlag, Berlin, 2005.

[Ka] D. A. Kazhdan, *On a connection between the dual space of a group and the structure of its closed subgroups*. Funct. Anal. Appl. **1** (1967), 63–65.

[KS] D. Kelmer and P. Sarnak, *Spectral gap for products of* PSL(2, *R*). arXiv:0808.2368.

[KM] D. Kleinbock and G. A. Margulis, *Logarithm laws for flows on homogeneous spaces*. Invent. Math. **138** (1999), 451–494.

[Kn] A. W. Knapp, *Representation Theory of Semisimple Groups: An Overview Based on Examples*. Princeton Math. Ser. **36**, Princeton University Press, Princeton, NJ, 1986.

[L]   J.-S. Li, *The minimal decay of matrix coefficients for classical groups*, Har-monic Analysis in China. Math. Appl. **327**, Kluwer, Dordrecht, 1995, pp. 146–169.

[LZ] J.-S. Li and C.B. Zhu, *On the decay of matrix coefficients of exceptional groups*. Math. Ann. **305** (1996), 249–270.

[LP] P. Lax and R. Phillips, *The asymptotic distribution of lattice points in Eu-clidean and non-Euclidean spaces*. J. Funct. Anal. **46** (1982), 280–350.

[M]   G. A. Margulis, *Discrete Subgroups of Semisimple Lie Groups*. Modern Sur-veys in Math. **17**, Springer Verlag, New York, 1991.

[MNS] G. Margulis, A. Nevo, and E. M. Stein, *Analogs of Wiener's ergodic theo-rems for semisimple Lie groups II*. Duke. Math. J. **103** (2000), 233–259.

[Ma] F. Maucourant, *Homogeneous asymptotic limits of Haar measures of semisimple linear groups and their lattices*. Duke Math. J. **136** (2007), 357–399.

[N1] A. Nevo, *Harmonic analysis and pointwise ergodic theorems for non-commuting transformations*. J. Amer. Math. Soc. **7** (1994), 875–902.

[N2] A. Nevo, *Pointwise ergodic theorems for radial averages on simple Lie groups I*. Duke Math. J. **76** (1994), 113–140.

[N3] A. Nevo, *Pointwise ergodic theorems for radial averages on simple Lie groups II*. Duke Math. J. **86** (1997), 239–259.

[N4] A. Nevo, *Spectral transfer and pointwise ergodic theorems for semisimple Kazhdan groups*. Math. Res. Lett. **5** (1998), 305–325.

[N5]   A. Nevo, *Exponential volume growth, maximal functions on symmetric spaces, and ergodic theorems for semisimple Lie groups*. Ergodic Theory Dynam. Systems **25** (2005), 1257–1294.

[N6]   A. Nevo, *Pointwise ergodic theorems for actions of groups*. In: Handbook of Dynamical Systems, vol. IB, Eds. B. Hasselblatt and A. Katok, 2006, Elsevier, Amsterdam, pp. 871–982.

[NeSa] A. Nevo and P. Sarnak, *Prime and almost prime integral points on principal homogeneous spaces*. arXiv:0902.0692.

[NS1] A. Nevo and E. M. Stein, *A generalization of Birkhoff pointwise ergodic theorem*. Acta Math. **173** (1994), 135–154.

[NS2] A. Nevo and E. M. Stein, *Analogs of Wiener's ergodic theorems for semisimple groups I*. Ann. Math. **145** (1997), 565–595.

[O]   H. Oh, *Uniform pointwise bounds for matrix coefficients of unitary representations and applications to Kazhdan constants*. Duke J. Math. **113** (2002), 133–192.

[R]   W. Rudin, *Real and Complex Analysis*, 2nd Edition, McGraw-Hill, New York, 1980.

[Se]   A. Selberg, *Harmonic analysis and discontinuous groups in weakly symmetric Riemannian spaces with applications to Dirichlet series*. J. Indian Math. Soc. **20** (1956), 47–87.

[Si]   A. J. Silberger, *Introduction to harmonic analysis on reductive p-adic groups*. Math. Notes **23**, Princeton University Press, Princeton, NJ, 1979.

[St]   G. Stuck, *Cocycles of ergodic group actions and vanishing of first cohomology for S-arithmetic groups*. Amer. J. Math. **113** (1991), 1–23.

[T]   J. Tits, *Reductive groups over local fields*. In: Automorphic Forms, Representations and L-functions, vol. I, Proc. Sympos. Pure Math. **33**, American Mathematical Society, Providence, RI, 1979, pp. 29–69.

[Ve]   A. Veca, *The Kunze-Stein phenomenon*. Ph. D. Thesis, University of New South Wales, 2002.

[Va]   V. S. Varadarajan, *Lie Groups, Lie Algebras, and Their Representations*. Graduate Texts in Math. **102**, Springer Verlag, New York, 1984.

[Z]   R. J. Zimmer, *Ergodic Theory and Semisimple Groups*, Birkhauser, Boston, 1984.

Index

Milton Keynes UK
Ingram Content Group UK Ltd.
UKHW020145080724
445185UK00008B/322

9 780691 141855